T0213476

SpringerBriefs in Mathematical Physics

Volume 36

SpringerBriefs are characterized in general by their size (50–125 pages) and fast production time (2–3 months compared to 6 months for a monograph).

Briefs are available in print but are intended as a primarily electronic publication to be included in Springer's e-book package.

Typical works might include:

- An extended survey of a field
- A link between new research papers published in journal articles
- A presentation of core concepts that doctoral students must understand in order to make independent contributions
- Lecture notes making a specialist topic accessible for non-specialist readers.

SpringerBriefs in Mathematical Physics showcase, in a compact format, topics of current relevance in the field of mathematical physics. Published titles will encompass all areas of theoretical and mathematical physics. This series is intended for mathematicians, physicists, and other scientists, as well as doctoral students in related areas.

Springer Briefs in a nutshell

SpringerBriefs specifications vary depending on the title. In general, each Brief will have:

- 50–125 published pages, including all tables, figures, and references
- Softcover binding
- Copyright to remain in author's name
- Versions in print, eBook, and MyCopy

More information about this series at http://www.springer.com/series/11953

Kohtaro Tadaki

A Statistical Mechanical Interpretation of Algorithmic Information Theory

Springer

Kohtaro Tadaki
Department of Computer Science
Chubu University
Kasugai, Japan

ISSN 2197-1757 ISSN 2197-1765 (electronic)
SpringerBriefs in Mathematical Physics
ISBN 978-981-15-0738-0 ISBN 978-981-15-0739-7 (eBook)
https://doi.org/10.1007/978-981-15-0739-7

This Springer imprint is published by the registered company Springer Nature Singapore Pte Ltd.
The registered company address is: 152 Beach Road, #21-01/04 Gateway East, Singapore 189721, Singapore

Preface

Algorithmic information theory (AIT, for short) is a field of mathematics. It is also known as algorithmic randomness, recently. AIT originated in the groundbreaking works of Solomonoff [36], Kolmogorov [18], and Chaitin [8] in the mid-1960s. They independently introduced the notion of program-size complexity, also known as Kolmogorov complexity, in order to quantify the randomness of an individual object. Around the same time, Martin-Löf [24] introduced a measure-theoretic approach to characterize the randomness of an individual infinite binary sequence, called Martin-Löf randomness nowadays. Later on, in the 1970s, Schnorr [33] and Chaitin [9] showed that Martin-Löf randomness is equivalent to the randomness defined by program-size complexity in characterizing random infinite binary sequences. In the twenty-first century, AIT makes remarkable progress through close interaction with recursion theory, i.e., computability theory. See Nies [27] and Downey and Hirschfeldt [14] for the recent development as well as the historical detail of AIT.

In this book, we develop a statistical mechanical interpretation of AIT, a new aspect of the theory. We reveal a close and coherent relationship between AIT and statistical mechanics by uncovering a computation-theoretic meaning of the notion of temperature in statistical mechanics. Specifically, in the statistical mechanical interpretation of AIT, the temperature has the meaning of the compression rate of the values of all the thermodynamic quantities by means of program-size complexity. This situation holds even for the temperature itself as a thermodynamic quantity. Actually, we have the situation that the compression rate of the temperature equals to the temperature itself. In the book, the various related results are presented and studied, based on the thermodynamic quantities of AIT.

In our opening chapter, we begin with a development of a statistical mechanical interpretation of noiseless source coding in information theory. The source coding problem is closely related to the framework of AIT. This chapter introduces basic ideas and intuitive notion useful for an intuitive understanding of the statistical mechanical interpretation of AIT developed through the subsequent chapters.

In Chap. 2, we review the basic definitions and results in AIT, where we introduce the notion of program-size complexity and present several but equivalent ways for defining the notion of randomness for an infinite binary sequence, i.e., for a real. In Chap. 3, we introduce the notion of partial randomness, which is a stronger representation of the compression rate by means of program-size complexity.

We start to develop the statistical mechanical interpretation of AIT in Chap. 4 by introducing the notion of thermodynamic quantities such as energy, free energy, entropy, and specific heat into AIT. We then investigate their properties. We see that the temperature equals to the partial randomness of the values of all these thermodynamic quantities. In statistical mechanics or thermodynamics, among all thermodynamic quantities, one of the most typical thermodynamic quantities is temperature itself. In Chap. 5, we see that the result of the previous chapter holds for temperature itself. That is, we obtain fixed point theorems on partial randomness, which state that the partial randomness of the temperature equals to the temperature itself.

In Chap. 6, based on a physical argument, we develop a total statistical mechanical interpretation of AIT which actualizes a perfect correspondence to normal statistical mechanics. We do this by identifying a microcanonical ensemble in the framework of AIT.

The values of all the thermodynamic quantities of AIT diverge when the temperature T exceeds 1. This phenomenon might be regarded as some sort of phase transition in statistical mechanics. In Chap. 7, we prove a theorem useful for analyzing the thermodynamic quantities of AIT and in Chap. 8 using this theorem, we clarify the computation-theoretic meaning of the phase transition.

Our closing chapter reviews the recent developments of the statistical mechanical interpretation of AIT, and then concludes the book with a mention of the future direction of this work.

All the arguments in this book are made with full mathematical rigor, except for ones made in Chap. 1, Sect. 4.2, and Chap. 6. In these exceptional parts, we make an argument on the same level of mathematical strictness as statistical mechanics in physics, with appealing to intuition. The rigorous parts of the book, which account for almost all of it, never use the results from these "physical" parts. The reader who is not interested in these physical but non-strict parts may skip them.

As a prerequisite for reading (the mathematical parts of) this book, we assume that the reader is familiar with the results and proofs in Chaitin [9]. In order to understand Chaitin [9], in turn, it would be enough to read one of the excellent textbooks of the theory of computation, including Sipser [35] or Hopcroft et al. [17].

I would like to express my special thanks to Atsuo Kuniba, a series editor of SpringerBriefs in Mathematical Physics, for inviting me to write this book.

This work was partially done with the support of each of JSPS KAKENHI Grant Numbers 20540134, 23650001, 23340020, and 24540142, of SCOPE from the Ministry of Internal Affairs and Communications of Japan, of CREST from Japan Science and Technology Agency, and of the Ministry of Economy, Trade and Industry of Japan.

I thank the staff of Springer, especially, Masayuki Nakamura, Yoshio Saito, and Chino Hasebe, for valuable assistance during the preparation of this book.

Kasugai, Japan   Kohtaro Tadaki

Contents

Chapter 1
Statistical Mechanical Interpretation of Noiseless Source Coding

In this chapter we develop a statistical mechanical interpretation of the noiseless source coding scheme based on an absolutely optimal instantaneous code. The notions in statistical mechanics such as statistical mechanical entropy, temperature, and thermal equilibrium are translated into the context of noiseless source coding. Especially, it is discovered that the temperature 1 corresponds to the average code-word length of an instantaneous code in this statistical mechanical interpretation of noiseless source coding scheme. This correspondence is also verified by the investigation using box-counting dimension. Using the notion of temperature and statistical mechanical arguments, some information-theoretic relations can be derived in the manner which appeals to intuition.

The statistical mechanical interpretation of the noiseless source coding scheme is a subject worth studying on its own, but the notion and ideas introduced and considered in the subject give a crucial intuition to a rigorous treatment of the statistical mechanical interpretation of AIT developed through the subsequent chapters.

1.1 Introduction

Source coding is one of the major subjects in information theory (Shannon [34], Ash [1], Cover and Thomas [13]). The problem of source coding is to study how to compress data, such as a text file on computer, into a binary string as small as possible. In this chapter, according to Tadaki [43], we develop a statistical mechanical interpretation of the *noiseless source coding scheme based on an instantaneous code*, which is the most basic source coding scheme in information theory. In particular, we introduce a statistical mechanical interpretation to the noiseless source coding scheme based on an *absolutely optimal* instantaneous code. Then the notions in

---

This chapter is an extended version of Tadaki [43].

© The Author(s), under exclusive license to Springer Nature Singapore Pte Ltd. 2019
K. Tadaki, *A Statistical Mechanical Interpretation of Algorithmic Information Theory*,
SpringerBriefs in Mathematical Physics 36,
https://doi.org/10.1007/978-981-15-0739-7_1

statistical mechanics, such as statistical mechanical entropy, temperature, and thermal equilibrium, are translated into the context of noiseless source coding.

We identify a coded message by an instantaneous code with an energy eigenstate of a quantum system treated in statistical mechanics, and the length of the coded message with the energy of the eigenstate. The discreteness of the length of coded message naturally corresponds to the statistical mechanics based on quantum mechanics and not on classical mechanics. This is because, in general, the energy of a system takes discrete value in quantum mechanics while that takes continuous value in classical mechanics. Especially, in this statistical mechanical interpretation of noiseless source coding, the energy of the corresponding quantum system to an instantaneous code is bounded to the above, and therefore the system has negative temperature, as in statistical mechanics. We discover that the temperature 1 corresponds to the average codeword length of an instantaneous code in the interpretation.

Note that *we do not stick to the mathematical strictness of the argument in this chapter*. We respect the statistical mechanical intuition in order to shed light on a hidden statistical mechanical aspect of information theory, and therefore make an argument on the same level of mathematical strictness as statistical mechanics.

1.2   Instantaneous Codes

We start with some notation on instantaneous codes from information theory (Shannon [34], Ash [1], Cover and Thomas [13]).

For any set S, we denote by $\#S$ the number of elements in S. We denote the set of all finite binary strings by $\{0, 1\}^*$. For any $s \in \{0, 1\}^*$, the *length* of s is denoted by $|s|$. We define an *alphabet* to be any nonempty finite set.

Let X be an arbitrary *random variable* with an alphabet χ and a *probability mass function* $p_X(x) = \Pr\{X = x\}$, $x \in \chi$. Then the *entropy* $H(X)$ of X is defined by

$$H(X) := -\sum_{x \in \chi} p_X(x) \log p_X(x),$$

where the log is to the base 2 (Shannon [34]). We will introduce the notion of a *statistical mechanical entropy* later. Thus, in order to distinguish $H(X)$ from it, we particularly call $H(X)$ the *Shannon entropy* of X. A subset S of $\{0, 1\}^*$ is called a *prefix-free set* if no string in S is a prefix of any other string in S. An *instantaneous code* C for the random variable X is an injective mapping from χ to $\{0, 1\}^*$ such that $C(\chi) := \{C(x) | x \in \chi\}$ is a prefix-free set. For each $x \in \chi$, We call $C(x)$ the *codeword* corresponding to x and denote $|C(x)|$ by $l(x)$. A sequence $x_1, x_2, \ldots, x_N$ with $x_i \in \chi$ is called a *message*. On the other hand, the finite binary string $C(x_1)C(x_2) \cdots C(x_N)$ is called the *coded message* for a message $x_1, x_2, \ldots, x_N$.

An instantaneous code plays a basic role in the noiseless source coding problem, as described as follows: Let $X_1, X_2, \ldots, X_N$ be *independent and identically distributed* random variables drawn from the probability mass function $p_X(x)$. The objective of the noiseless source coding problem is to minimize the length of the binary string $C(x_1)C(x_2)\cdots C(x_N)$ for a message $x_1, x_2, \ldots, x_N$ generated by the random variables $\{X_i\}$ as $N \to \infty$. For that purpose, it is sufficient to consider the *average codeword length* $L_X(C)$ of the instantaneous code C for the random variable X, which is defined by

$$L_X(C) := \sum_{x \in \chi} p_X(x) l(x)$$

independently on the value of N. We can then show that

$$L_X(C) \geq H(X)$$

for any instantaneous code C for the random variable X (Shannon [34]). Hence, the Shannon entropy gives the *data compression limit* for the noiseless source coding problem based on instantaneous codes. Thus, it is important to consider the notion of *absolutely optimality* of an instantaneous code, where we say that an instantaneous code C for the random variable X is *absolutely optimal* if $L_X(C) = H(X)$. We can then show the following theorem.

Theorem 1.1 *An instantaneous code C for a random variable X with an alphabet χ is absolutely optimal if and only if $p_X(x) = 2^{-l(x)}$ for all $x \in \chi$.* $\square$

Finally, for each $x^N = (x_1, x_2, \ldots, x_N) \in \chi^N$, we define $p_X(x^N)$ as

$$p_X(x_1) p_X(x_2) \cdots p_X(x_N).$$

In this chapter, we develop a statistical mechanical interpretation of the noiseless source coding by means of an instantaneous code. In particular, we assume that an instantaneous code C for a random variable X is *absolutely optimal* throughout the rest of this chapter.

1.3 Equilibrium Statistical Mechanics

We start with some reviews of equilibrium statistical mechanics (Toda et al. [57], Callen [2]).

In statistical mechanics, we consider a quantum system $\mathscr{S}_{\text{total}}$ which consists in a large number of identical quantum subsystems. Let N be the number of such subsystems. For example, $N \sim 10^{22}$ for $1\,\text{cm}^3$ of a gas at room temperature. We assume here that each quantum subsystem can be distinguishable from others. Thus, we deal with quantum particles which obey *Maxwell-Boltzmann statistics* and neither Bose-Einstein statistics nor Fermi-Dirac statistics. Under this assumption, we can identify the ith quantum subsystem $\mathscr{S}_i$ for each $i = 1, \ldots, N$. In quantum mechanics, any quantum system is described by a quantum state *completely*. In statistical mechanics, among all quantum states, *energy eigenstates* are of particular importance. Any energy eigenstate of each subsystem $\mathscr{S}_i$ can be specified by a number $n = 1, 2, 3, \ldots$, called a *quantum number*, where the subsystem in the energy eigenstate specified by n has the energy E_n. Then, any energy eigenstate of the system $\mathscr{S}_{\text{total}}$ can be specified by an N-tuple $(n_1, n_2, \ldots, n_N)$ of quantum numbers. If the state of the system $\mathscr{S}_{\text{total}}$ is the energy eigenstate specified by $(n_1, n_2, \ldots, n_N)$, then the state of each subsystem $\mathscr{S}_i$ is the energy eigenstate specified by n_i and the system $\mathscr{S}_{\text{total}}$ has the energy $E_{n_1} + E_{n_2} + \cdots + E_{n_N}$. Then, the fundamental postulate of statistical mechanics, called the *principle of equal probability*, is stated as follows.

The Principle of Equal Probability: If the energy of the system $\mathscr{S}_{\text{total}}$ is known to have a constant value in the range between E and $E + \delta E$, where δE is the indeterminacy in measurement of the energy of the system $\mathscr{S}_{\text{total}}$, then the system $\mathscr{S}_{\text{total}}$ is equally likely to be in any energy eigenstate specified by $(n_1, n_2, \ldots, n_N)$ such that $E \leq E_{n_1} + E_{n_2} + \cdots + E_{n_N} \leq E + \delta E$.   □

Let $\Theta(E, N)$ be the total number of energy eigenstates of $\mathscr{S}_{\text{total}}$ specified by $(n_1, n_2, \ldots, n_N)$ such that $E \leq E_{n_1} + E_{n_2} + \cdots + E_{n_N} \leq E + \delta E$. The above postulate states that any energy eigenstate of $\mathscr{S}_{\text{total}}$ whose energy lies between E and $E + \delta E$ occurs with the probability $1/\Theta(E, N)$. This uniform distribution of energy eigenstates whose energy lies between E and $E + \delta E$ is called a *microcanonical ensemble*. In statistical mechanics, the *entropy* $S(E, N)$ of the system $\mathscr{S}_{\text{total}}$ is then defined by

$$S(E, N) := k_{\text{B}} \ln \Theta(E, N),$$

where k_{B} is a positive constant, called the *Boltzmann constant*, and the ln denotes the natural logarithm. The *average energy* ε per one subsystem is given by E/N. In a normal case where ε has a finite value, the entropy $S(E, N)$ is proportional to N. On the other hand, the indeterminacy δE of the energy contributes to $S(E, N)$ through the term $k_{\text{B}} \ln \delta E$, which can be ignored compared to N unless δE is too small. Thus, the magnitude of the indeterminacy δE of the energy does not matter to the value of the entropy $S(E, N)$ unless it is too small. The *temperature* $T(E, N)$ of the system $\mathscr{S}_{\text{total}}$ is defined by

$$\frac{1}{T(E, N)} := \frac{\partial S}{\partial E}(E, N).$$

Thus, the temperature is a function of E and N.

1.4   Statistical Mechanical Interpretation

Now we give a statistical mechanical interpretation to the noiseless source coding scheme based on an instantaneous code. Let X be an arbitrary random variable with an alphabet χ, and let C be an absolutely optimal instantaneous code for the random variable X. Let $X_1, X_2, \ldots, X_N$ be independent and identically distributed random variables drawn from the probability mass function $p_X(x)$ for a large N, say $N \sim 10^{22}$. We relate the noiseless source coding based on C to the equilibrium statistical mechanics developed in the preceding section, as follows.

The sequence $X_1, X_2, \ldots, X_N$ corresponds to the quantum system $\mathscr{S}_{\text{total}}$, where each X_i corresponds to the ith quantum subsystem $\mathscr{S}_i$. We relate $x \in \chi$, or equivalently a codeword $C(x)$, to an energy eigenstate of a subsystem specified by n, and we relate the length $|C(x)|$ of the codeword $C(x)$ to the energy E_n of the energy eigenstate of the subsystem specified by n. Then, a sequence $(x_1, \ldots, x_N) \in \chi^N$, or equivalently a concatenation $C(x_1) \cdots C(x_N)$, corresponds to an energy eigenstate of $\mathscr{S}_{\text{total}}$ specified by $(n_1, \ldots, n_N)$, and the length $|C(x_1) \cdots C(x_N)|$ of the concatenation $C(x_1) \cdots C(x_N)$ corresponds to the energy $E_{n_1} + \cdots + E_{n_N}$ of the energy eigenstate of $\mathscr{S}_{\text{total}}$ specified by $(n_1, \ldots, n_N)$.

Thus, in our statistical mechanical interpretation, *the notion of concatenation of codewords corresponds to the notion of tensor product of quantum states in quantum mechanics.* Actually, in the above setting, the concatenation $C(x_1) \cdots C(x_N)$ of codewords $C(x_1), \ldots, C(x_N)$ corresponds to the tensor product of N energy eigenstates, to which the codewords $C(x_1), \ldots, C(x_N)$ correspond, respectively. Moreover, in our statistical mechanical interpretation, *the length of a codeword or the concatenation of codewords corresponds to the energy of the corresponding energy eigenstate.* Note the requirement of statistical mechanics that if the quantum subsystems $\mathscr{S}_1, \ldots, \mathscr{S}_N$ have energy $E_{n_1}, \ldots, E_{n_N}$, respectively, then the total quantum system $\mathscr{S}_{\text{total}}$ has the energy $E_{n_1} + \cdots + E_{n_N}$. In other words, the *additivity of energy* must hold in statistical mechanics. This requirement is *automatically* satisfied in our statistical mechanical interpretation, since the equality

$$|C(x_1) \cdots C(x_N)| = |C(x_1)| + \cdots + |C(x_N)|$$

always holds due to the definition of concatenation, where $|C(x_1) \cdots C(x_N)|$ is interpreted as an energy of the total system while each $|C(x_i)|$ is interpreted as an energy of the ith subsystem. Thus, our statistical mechanical interpretation of an instantaneous code is natural and consistent.

We define a subset $C(L, N)$ of $\{0, 1\}^*$ as the set of all coded messages

$$C(x_1) \cdots C(x_N)$$

whose length lies between L and $L + \delta L$. Then $\Theta(L, N)$ is defined as $\#C(L, N)$. Therefore $\Theta(L, N)$ is the total number of coded messages whose length lies between L and $L + \delta L$. Since C is an absolutely optimal instantaneous code for X, it follows from Theorem 1.1 that if $C(x_1) \cdots C(x_N) \in C(L, N)$ then

$$2^{-(L+\delta L)} \le p_X(x^N) \le 2^{-L}.$$

Thus, all coded messages $C(x_1) \cdots C(x_N) \in C(L, N)$ occur with the probability 2^{-L}. Note here that we care nothing about the magnitude of δL, as in the case of statistical mechanics. Thus, the following principle, called the *principle of equal conditional probability*, holds.

The Principle of Equal Conditional Probability: Given that the length of coded message is L, all coded messages occur with the same probability $1/\Theta(L, N)$.   □

We identify a *microcanonical ensemble* in the noiseless source coding scheme in this manner. Thus, we can develop a certain sort of equilibrium statistical mechanics on the noiseless source coding scheme. This is possible because, in order to do so, we only have to follow the theoretical development of normal equilibrium statistical mechanics, starting from this microcanonical ensemble.

Note that, in statistical mechanics, the principle of equal probability is just a conjecture which is not yet proved completely in a realistic physical system. On the other hand, in our statistical mechanical interpretation of noiseless source coding, the principle of equal conditional probability is *automatically* satisfied without any assumption, due to the *intrinsic probabilistic nature* of the information source, i.e., the independent and identically distributed random variables $\{X_i\}$.

The *statistical mechanical entropy* $S(L, N)$ of the instantaneous code C is defined by

$$S(L, N) := \log \Theta(L, N). \tag{1.1}$$

The *temperature* $T(L, N)$ of C is then defined by

$$\frac{1}{T(L, N)} := \frac{\partial S}{\partial L}(L, N). \tag{1.2}$$

Thus, the temperature is a function of L and N. The *average length* λ of coded message per one codeword is given by L/N. The average length λ corresponds to the average energy ε in the statistical mechanics reviewed in the previous section.

1.5   Properties of Statistical Mechanical Entropy

In statistical mechanics, it is important to know the values of the energy E_n of a subsystem $\mathscr{S}_i$ for all quantum numbers n, since the values determine the entropy $S(E, N)$ of the quantum system $\mathscr{S}_{\text{total}}$. Corresponding to this fact, the knowledge of $l(x)$ for all $x \in \chi$ is important to calculate $S(L, N)$. We investigate some properties of $S(L, N)$ and $T(L, N)$ based on $l(x)$ in the following.

As is well known in statistical mechanics, if the energy of a quantum system $\mathscr{S}_{\text{total}}$ is bounded to the above, then the system can have negative temperature. The same situation happens in our statistical mechanics developed on an instantaneous code C, since there are only finite codewords of C. We define $l_{\min}$ and $l_{\max}$ as $\min\{l(x) \mid x \in \chi\}$ and $\max\{l(x) \mid x \in \chi\}$, respectively. Given N, the statistical mechanical entropy $S(L, N)$ is a unimodal function of L and takes nonzero value only between $Nl_{\min}$ and $Nl_{\max}$. Let L_0 be the value L which maximizes $S(L, N)$. If $L < L_0$ then $T(L, N) > 0$.

On the other hand, if $L > L_0$ then $T(L, N) < 0$. The temperature $T(L, N)$ takes $\pm\infty$ at $L = L_0$.

According to the method of Boltzmann and Planck (see e.g. Toda et al. [57]), we can show that

$$S(L, N) = NH(R(C; L, N)). \tag{1.3}$$

Here $R(C; L, N)$ is a random variable with the alphabet χ and the probability mass function $p_{R(C;L,N)}(x) = \Pr\{R(C; L, N) = x\}$ defined by

$$p_{R(C;L,N)}(x) := \frac{2^{-\beta(L,N)l(x)}}{\sum_{a \in \chi} 2^{-\beta(L,N)l(a)}} \tag{1.4}$$

with some function $\beta(L, N)$ of L and N. This function $\beta(L, N)$ is determined by the condition

$$\frac{L}{N} = \sum_{x \in \chi} l(x) p_{R(C;L,N)}(x). \tag{1.5}$$

Let us calculate the form of the function $\beta(L, N)$. First, using (1.3), (1.4), and (1.5) we have

$$S(L, N) = \beta(L, N)L + N \log \sum_{a \in \chi} 2^{-\beta(L,N)l(a)}.$$

Differentiating this equation on L and then applying (1.5) to the result, we have

$$\frac{\partial S}{\partial L}(L, N) = \beta(L, N).$$

It follows from (1.2) that

$$\beta(L, N) = \frac{1}{T(L, N)}.$$

Hence, in summary, the statistical mechanical entropy $S(L, N)$ is given by

$$S(L, N) = NH(G(C; T(L, N))), \tag{1.6}$$

where $G(C; T)$ is a random variable with the alphabet χ and the probability mass function $p_{G(C;T)}(x) = \Pr\{G(C; T) = x\}$ defined by

$$p_{G(C;T)}(x) := \frac{2^{-l(x)/T}}{\sum_{a \in \chi} 2^{-l(a)/T}}. \tag{1.7}$$

The temperature $T(L, N)$ is implicitly determined through the equation

$$\frac{L}{N} = \sum_{x \in \chi} l(x) p_{G(C;T(L,N))}(x) \tag{1.8}$$

as a function of L/N, i.e., the average length λ. Note that these properties of $S(L, N)$ and $T(L, N)$ are derived only based on a *combinatorial aspect* of $S(L, N)$.

Now, let us take into account the *probabilistic issue* given by the random variables $X_1, X_2, \ldots, X_N$. Since the instantaneous code C is absolutely optimal for X, it follows from Theorem 1.1 that any particular coded message of length L occurs with probability 2^{-L}. Thus, the probability that some coded message of length L occurs is given by $2^{-L}\Theta(L, N)$. Hence, by differentiating $2^{-L}\Theta(L, N)$ on L and setting the result to 0, we can determine the most probable length L^* of coded message, given N. Thus, we have the relation

$$\frac{\partial}{\partial L}\{-L + S(L, N)\}\Big|_{(L,N)=(L^*,N)} = 0,$$

which is satisfied by L^*. It follows that $T(L^*, N) = 1$. Therefore, the temperature 1 corresponds to the most probable length L^*. On the other hand, using (1.7) we have that $p_{G(C;1)}(x) = 2^{-l(x)}$ at $T(L^*, N) = 1$, and therefore, by (1.8), we have $L^*/N = L_X(C)(= H(X))$. This result is consistent with the *law of large numbers*. Thus, at the temperature 1, the average length λ of coded message per one codeword coincides with the average codeword length $L_X(C)$. In this way the temperature 1 corresponds to the average codeword length $L_X(C)$.

1.6   Thermal Equilibrium Between Two Instantaneous Codes

In this section, we consider the notion of "thermal equilibrium" between two instantaneous codes. Let X^{I} be an arbitrary random variable with an alphabet χ^{I}, and let C^{I} be an absolutely optimal instantaneous code for the random variable X^{I}. Let $X_1^{\mathrm{I}}, X_2^{\mathrm{I}}, \ldots, X_{N^{\mathrm{I}}}^{\mathrm{I}}$ be independent and identically distributed random variables drawn from the probability mass function $p_{X^{\mathrm{I}}}(x)$ for a large N^{I}. On the other hand, let X^{II} be an arbitrary random variable with an alphabet χ^{II}, and let C^{II} be an absolutely optimal instantaneous code for the random variable X^{II}. Let $X_1^{\mathrm{II}}, X_2^{\mathrm{II}}, \ldots, X_{N^{\mathrm{II}}}^{\mathrm{II}}$ be independent and identically distributed random variables drawn from the probability mass function $p_{X^{\mathrm{II}}}(x)$ for a large N^{II}. Note that χ^{I} is not necessarily identical to χ^{II}, and C^{I} are not necessarily identical to C^{II}, as well. Moreover, we do not assume that $N^{\mathrm{I}} = N^{\mathrm{II}}$.

Consider the following problem: Find the most probable values L^{I} and L^{II}, given that the sum $L^{\mathrm{I}} + L^{\mathrm{II}}$ of the length L^{I} of coded message by C^{I} for the random variables $\{X_i^{\mathrm{I}}\}$ and the length L^{II} of coded message by C^{II} for the random variables $\{X_j^{\mathrm{II}}\}$ is equal to L.

In order to solve this problem, the statistical mechanical notion of "thermal equilibrium" can be used. We first note that any particular coded message by C^{I} of length L_{I} and any particular coded message by C^{II} of length L_{II} simultaneously occur with probability $2^{-L_{\mathrm{I}}}2^{-L_{\mathrm{II}}} = 2^{-L}$, since the instantaneous codes C^{I} and C^{II} are absolutely optimal. Thus, any particular pair of coded messages by C^I and C^{II} occurs with an equal probability, given that the total length of coded messages for $\{X_i^I\}$ and $\{X_j^{II}\}$ is L. Therefore, the most probable allocation (L_I^*, L_{II}^*) of $L = L_I + L_{II}$ maximizes the product $\Theta_I(L_I, N_I)\Theta_{II}(L_{II}, N_{II})$. Using (1.1) and (1.2) we see that this condition is equivalent to the equality:

$$T_I(L_I^*, N_I) = T_{II}(L_{II}^*, N_{II}), \tag{1.9}$$

where the functions T_I and T_{II} are the temperature of C^I and C^{II}, respectively. This equality corresponds to the condition on the thermal equilibrium between two systems, given a total energy, in statistical mechanics.

Based on (1.9), the most probable allocation (L_I^*, L_{II}^*) is actually calculated as follows: First, the value of $T_I(L_I^*, N_I)(= T_{II}(L_{II}^*, N_{II}))$ is obtained by solving the equation on T:[1]

$$\frac{N^I}{L} \sum_{x \in \chi^I} \left|C^I(x)\right| p_{G(C^I;T)}(x) + \frac{N^{II}}{L} \sum_{x \in \chi^{II}} \left|C^{II}(x)\right| p_{G(C^{II};T)}(x) = 1. \tag{1.10}$$

This equation, satisfied by $T_I(L_I^*, N_I)$, can be derived from the equation $L = L_I^* + L_{II}^*$ using (1.8). Once the value of $T_I(L_I^*, N_I)$ is calculated, using (1.8) again, each of the most probable values L_I^* and L_{II}^* can be determined.

1.7 Dimension of Coded Messages

The notion of *dimension* plays an important role in *fractal geometry* (see e.g. Falconer [15]). In this section, we investigate our statistical mechanical interpretation of the noiseless source coding from the point of view of dimension.

Let F be a bounded subset of $\mathbb{R}$, and let $N_n(F)$ be the number of 2^{-n}-*mesh cubes* that intersect F, where 2^{-n}-mesh cube is a subset of $\mathbb{R}$ in the form of

$$[m2^{-n}, (m+1)2^{-n}]$$

for some integer m. The *box-counting dimension* $\dim_B F$ of F is then defined by

$$\dim_B F := \lim_{n \to \infty} \frac{\log N_n(F)}{n}.$$

Let $\{0, 1\}^\infty = \{b_1 b_2 b_3 \cdots \mid b_i = 0, 1 \text{ for all } i = 1, 2, 3, \dots\}$ be the set of all infinite binary strings. Tadaki [40, 41] investigated the dimension of sets of coded messages of infinite length, where the number of distinct codewords is finite or infinite. In a similar manner, we investigate the box-counting dimension of a set of coded messages

---

[1]In the actual calculation, we solve the Eq. (1.10) on $2^{-1/T}$ and not on T.

of infinite length by an arbitrary absolutely optimal instantaneous code C, where each coded message of infinite length corresponds to a specific temperature.

By (1.8), the ratio L/N is uniquely determined by temperature T. Thus, by letting $L, N \to \infty$ while keeping the ratio L/N constant, we can regard the set $C(L, N)$ as a subset of $\{0, 1\}^\infty$. This kind of limit is called the *thermodynamic limit* in statistical mechanics. Taking the thermodynamic limit, we denote $C(L, N)$ by $F(T)$, where T is related to the limit value of L/N through (1.8). Although $F(T)$ is a subset of $\{0, 1\}^\infty$, we can regard $F(T)$ as a subset of $[0, 1]$ by identifying $\alpha \in \{0, 1\}^\infty$ with the real number $0.\alpha$. In this manner, we can consider the box-counting dimension $\dim_B F(T)$ of $F(T)$.

In what follows, we investigate the dependency of $\dim_B F(T)$ on temperature T with $-\infty \leq T \leq \infty$. First, it can be shown that

$$\dim_B F(T) = \lim_{L,N \to \infty} \frac{\log \Theta(L, N)}{L} = \lim_{L,N \to \infty} \frac{S(L, N)}{L},$$

where the limits are taken while satisfying (1.8) for each T. Thus the statistical mechanical entropy $S(L, N)$ and the box-counting dimension $\dim_B F(T)$ of $F(T)$ are closely related. Using (1.6), (1.7), and (1.8) we obtain, as an explicit formula of T,

$$\dim_B F(T) = \frac{1}{T} + \frac{1}{\lambda(T)} \log \sum_{x \in \chi} 2^{-l(x)/T}, \tag{1.11}$$

where $\lambda(T)$ is defined by

$$\lambda(T) := \sum_{x \in \chi} l(x) p_{G(C;T)}(x).$$

We define the "degeneracy factors" $d_{\min}$ and $d_{\max}$ of the lowest and highest "energies" of the instantaneous code C by $d_{\min} := \#\{x \in \chi \mid l(x) = l_{\min}\}$ and $d_{\max} := \#\{x \in \chi \mid l(x) = l_{\max}\}$, respectively. Note here that $d_{\max}$ is shown to be an even number, since $\sum_{x \in \chi} 2^{-l(x)} = 1$ holds due to the absolute optimality of C. In the increasing order of the ratio L/N (i.e., $\lambda(T)$), we see from (1.11) that

$$\lim_{T \to +0} \dim_B F(T) = \frac{\log d_{\min}}{l_{\min}},$$

$$\dim_B F(1) = 1,$$

$$\lim_{T \to \pm\infty} \dim_B F(T) = \frac{n \log n}{\sum_{x \in \chi} l(x)},$$

$$\lim_{T \to -0} \dim_B F(T) = \frac{\log d_{\max}}{l_{\max}}.$$

We can show that $n \log n < \sum_{x \in \chi} l(x)$ unless all codewords of C have the same length, and obviously $\log d_{\min}/l_{\min} < 1$ and $\log d_{\max}/l_{\max} < 1$ except for such a trivial case. Thus, in general, the dimension $\dim_B F(T)$ is maximized at the temperature $T = 1$.

This fact can also be verified based on the differentiation of $\dim_B F(T)$ on T. Actually, using (1.11) we can show that, if all codewords of C do not have the same length, then the following hold:

(i) $\dfrac{d}{dT} \dim_B F(T) \bigg|_{T=T_0} = 0$ if and only if $T_0 = 1$,

(ii) $\dfrac{d^2}{dT^2} \dim_B F(T) \bigg|_{T=1} < 0$.

We have seen above that $\dim_B F(1) = 1$. Thus, $\dim_B F(1)$ equals to $\dim_B[0, 1]$. This means that the set $F(1)$ is as rich as the interval $[0, 1]$ in a certain sense. This result can be explained as follows: Since C is an absolutely optimal instantaneous code, the equality $\sum_{x \in \chi} 2^{-l(x)} = 1$ holds. Therefore, all coded messages $C(x_1)C(x_2) \cdots$ of infinite length form the set $\{0, 1\}^\infty$ and therefore form the interval $[0, 1]$. Moreover, since C is absolutely optimal again, each coded message of infinite length occurs according to the uniform measure on $[0, 1]$, i.e., Lebesgue measure on $[0, 1]$. On the other hand, by the law of large numbers, the length of coded message for a message of length N is likely to equal $N L_X(C)$ for a sufficiently large N. These observations show that coded messages of length $N L_X(C)$ fill the interval $[0, 1]$ in a certain sense for a sufficiently large N. Thus, since the temperature $T = 1$ corresponds to these coded messages of length $N L_X(C)$, as seen in Sect. 1.5, the set $F(T)$ at the temperature $T = 1$ (i.e., $F(1)$) is as rich as the interval $[0, 1]$ in a certain sense.

Thus, $F(1)$ consists of the coded messages for all the messages which form the *typical set* in a sense, and therefore $\dim_B F(1) = 1$ holds.

1.8 Toward a Statistical Mechanical Interpretation of AIT

In this chapter we have developed a statistical mechanical interpretation of the noiseless source coding scheme based on an absolutely optimal instantaneous code. The notions in statistical mechanics such as statistical mechanical entropy, temperature, and thermal equilibrium are translated into the context of information theory. Especially, it is discovered that the temperature 1 corresponds to the average codeword length $L_X(C)$ in this statistical mechanical interpretation of information theory. This correspondence is also verified by the investigation using box-counting dimension. The argument is not necessarily mathematically rigorous. However, using the notion of temperature and statistical mechanical arguments, several information-theoretic relations can be derived in the manner which appeals to intuition.

From the next chapter, we are developing a statistical mechanical interpretation of AIT. As will be explained, one of the important concepts in AIT is the notion of *prefix-free machine*, which is used to define the notion of *program-size complexity*. The codewords of an instantaneous code form a finite prefix-free set of finite binary strings, and thus an instantaneous code can be regarded as a prefix-free machine M such that the domain of the definition of M is a finite set and M is injection. Conversely, in the context of source coding, a prefix-free machine can be regarded as a decoding equipment. Hence, the notion of prefix-free machine is a generalization of the notion of instantaneous code, where the *computability issue* is *newly* and *naturally* introduced into the definition. Therefore, the concepts introduced in this chapter from equilibrium statistical mechanics, such as statistical mechanical entropy, temperature, and thermal equilibrium, can be *naturally* transferred into the context of AIT.

In the statistical mechanical interpretation of AIT, however, the introduction of the computability issue brings a *new aspect* into the framework. We will see that the temperature has the meaning of the *compression rate* of the values of all the thermodynamic quantities of AIT *by means of program-size complexity*. This situation holds even for the temperature itself as a thermodynamic quantity, which results in the *fixed point theorems on partial randomness*.

This chapter has served as a "physical" introduction to the book. The notions and ideas which have been introduced and then studied in this chapter give an intuition to a rigorous treatment of the statistical mechanical interpretation of AIT developed through the rest of the book. In this chapter, we have not stuck to the mathematical strictness of the argument, and have made an argument on the same level of mathematical strictness as statistical mechanics. In the rest of the book, however, we make all the arguments with full mathematical rigor, except for the arguments made in Sect. 4.2 and Chap. 6, which motivate and explain the definitions of the thermodynamic quantities of AIT, introduced and studied in the rigorous parts of the book.

Chapter 2
Algorithmic Information Theory

In this chapter, we review the basic framework of algorithmic information theory to the extent necessary to read the rest of the book.

2.1 Basic Notation and Definitions

We start with some notation about numbers and strings which will be used in the rest of the book. $\mathbb{N} = \{0, 1, 2, 3, \dots\}$ is the set of natural numbers, and $\mathbb{N}^+$ is the set of positive integers. $\mathbb{Z}$ is the set of integers, and $\mathbb{Q}$ is the set of rationals. $\mathbb{R}$ is the set of reals.

$$\{0, 1\}^* = \{\lambda, 0, 1, 00, 01, 10, 11, 000, 001, 010, \dots\}$$

is the set of finite binary strings where λ denotes the *empty string*, and $\{0, 1\}^*$ is ordered as indicated. We identify any string in $\{0, 1\}^*$ with a natural number in this order, i.e., we consider $\varphi \colon \{0, 1\}^* \to \mathbb{N}$ such that $\varphi(s) = 1s - 1$ where the concatenation $1s$ of strings 1 and s is regarded as a dyadic integer, and then we identify s with $\varphi(s)$. For any $s \in \{0, 1\}^*$, $|s|$ is the *length* of s. For any $n \in \mathbb{N}$, we denote by $\{0, 1\}^n$ the set $\{s \mid s \in \{0, 1\}^* \ \& \ |s| = n\}$. A subset S of $\{0, 1\}^*$ is called *prefix-free* if no string in S is a prefix of another string in S. The so-called *Kraft inequality*

$$\sum_{s \in S} 2^{-|s|} \leq 1 \tag{2.1}$$

holds for every prefix-free set S.

$\{0, 1\}^\infty$ is the set of infinite binary sequences, where an infinite binary sequence is infinite to the right but finite to the left. Let $\omega \in \{0, 1\}^\infty$. For any $n \in \mathbb{N}$, we denote by $\omega\restriction_n$ the prefix of ω of length n. Thus, especially, $\omega\restriction_0$ is the empty string λ.

Let α be an arbitrary real. As usual, $\lfloor \alpha \rfloor$ denotes the greatest integer less than or equal to α, and $\lceil \alpha \rceil$ denotes the smallest integer greater than or equal to α. Note that there exists a unique $\zeta \in [0, 1)$ such that $\alpha - \zeta \in \mathbb{Z}$. We denote by

© The Author(s), under exclusive license to Springer Nature Singapore Pte Ltd. 2019
K. Tadaki, *A Statistical Mechanical Interpretation of Algorithmic Information Theory*,
SpringerBriefs in Mathematical Physics 36,
https://doi.org/10.1007/978-981-15-0739-7_2

$\mathrm{Binary_R}(\alpha) \in \{0, 1\}^\infty$ the base-two expansion of the real ζ with *infinitely many zeros*. Thus, we have $\alpha - \lfloor \alpha \rfloor = 0.(\mathrm{Binary_R}(\alpha))$ where the base-two expansion contains infinitely many zeros. Then, for any $n \in \mathbb{N}$, we denote $\mathrm{Binary_R}(\alpha)\lceil_n$ by $\alpha\lceil_n^R$. Hence, we have

$$0.(\alpha\lceil_n^R) \le \alpha - \lfloor \alpha \rfloor < 0.(\alpha\lceil_n^R) + 2^{-n} \qquad (2.2)$$

for every $n \in \mathbb{N}^+$.

On the other hand, note that there exists a unique $\eta \in (0, 1]$ such that $\alpha - \eta \in \mathbb{Z}$. We denote by $\mathrm{Binary_L}(\alpha) \in \{0, 1\}^\infty$ the base-two expansion of the real η with *infinitely many ones*. Thus, we have $\alpha - \lceil \alpha \rceil + 1 = 0.(\mathrm{Binary_L}(\alpha))$ where the base-two expansion contains infinitely many ones. Then, for any $n \in \mathbb{N}$, we denote $\mathrm{Binary_L}(\alpha)\lceil_n$ by $\alpha\lceil_n^L$. Hence, we have

$$0.(\alpha\lceil_n^L) < \alpha - \lceil \alpha \rceil + 1 \le 0.(\alpha\lceil_n^L) + 2^{-n} \qquad (2.3)$$

for every $n \in \mathbb{N}^+$.

A *dyadic rational* is a real of the form $m2^{-n}$ with $m \in \mathbb{Z}$ and $n \in \mathbb{N}$. It is then easy to see that α is a dyadic rational if and only if $\mathrm{Binary_R}(\alpha) \ne \mathrm{Binary_L}(\alpha)$.

We denote $\mathrm{Binary_R}(\alpha)$ by $\mathrm{Binary}\,(\alpha)$, and denote $\mathrm{Binary}\,(\alpha)\lceil_n$ by $\alpha\lceil_n$ for any $n \in \mathbb{N}$. Hence, *intuitively*, we identify a real α with the infinite binary sequence $\mathrm{Binary}\,(\alpha)$, i.e., the infinite binary sequence ω such that $0.\omega$ is the base-two expansion of $\alpha - \lfloor \alpha \rfloor$ with infinitely many zeros.

For any function f, the domain of definition of f is denoted by dom f.

Let $f: S \to \mathbb{R}$ with $S \subset \mathbb{R}$. We say that f is *increasing* (resp., *decreasing*) if $f(x) < f(y)$ (resp., $f(x) > f(y)$) for all $x, y \in S$ with $x < y$. On the other hand, we say that f is *non-increasing* (resp., *non-decreasing*) if $f(x) \ge f(y)$ (resp., $f(x) \le f(y)$) for all $x, y \in S$ with $x < y$. We denote by f' the derived function of f.

Normally, $o(n)$ denotes any function $f: \mathbb{N}^+ \to \mathbb{R}$ such that $\lim_{n \to \infty} f(n)/n = 0$. On the other hand, $O(1)$ denotes any function $g: \mathbb{N}^+ \to \mathbb{R}$ such that there is $C \in \mathbb{R}$ with the property that $|g(n)| \le C$ for all $n \in \mathbb{N}^+$.

2.2 Computability

A partial function $f: \mathbb{N}^+ \to \mathbb{N}^+$ is called *partial recursive* if there exists a deterministic Turing machine $\mathcal{M}$ such that, for each $n \in \mathbb{N}^+$, when executing $\mathcal{M}$ with the input n,

(i) if $n \in \mathrm{dom}\, f$ then the computation of $\mathcal{M}$ eventually terminates with output $f(n)$;
(ii) if $n \notin \mathrm{dom}\, f$ then the computation of $\mathcal{M}$ does not terminate.

A partial recursive function $f: \mathbb{N}^+ \to \mathbb{N}^+$ is called *total recursive* if dom $f = \mathbb{N}^+$. A total recursive function is also called a *recursive function* or *computable function*.

A subset S of $\{0, 1\}^*$ is called *recursively enumerable* (*r.e.*, for short) if there exists a deterministic Turing machine $\mathcal{M}$ such that, for each $x \in \{0, 1\}^*$, when executing $\mathcal{M}$ with the input x,

(i) if $x \in S$ then the computation of $\mathcal{M}$ eventually terminates;
(ii) if $x \notin S$ then the computation of $\mathcal{M}$ does not terminate.

The recursive enumerability of a subset of $\mathbb{N}^+$, $\mathbb{N}$, or $\mathbb{N}^+ \times \{0, 1\}^*$ is defined in a similar manner. See Sipser [35] or Hopcroft et al. [17] for the basic definitions and results of the theory of computation.

A sequence $\{a_n\}_{n \in \mathbb{N}}$ of numbers (rationals or reals) is called *increasing* if $a_{n+1} > a_n$ for all $n \in \mathbb{N}$. A real α is called *recursively enumerable* (*r.e.*, for short) if there exists a computable, increasing sequence of rationals which converges to α. An r.e. real is also called a *left-computable* real. On the other hand, we say that a real α is *right-computable* if $-\alpha$ is left-computable.

Let α and β be arbitrary r.e. reals. Then $\alpha + \beta$ is r.e. If α and β are non-negative, then $\alpha\beta$ is r.e.

It is easy to see that, for every real α, the following conditions are equivalent:

(i) α is r.e.
(ii) There exists a total recursive function $f : \mathbb{N}^+ \to \mathbb{Q}$ such that $f(n) \leq \alpha$ for all $n \in \mathbb{N}^+$ and $\lim_{n \to \infty} f(n) = \alpha$.[1]
(iii) The set $\{r \in \mathbb{Q} \mid r < \alpha\}$ is r.e.

An infinite binary sequence ω is *computable* if the mapping $\mathbb{N}^+ \ni n \mapsto \omega\!\restriction_n$ is a total recursive function. A real α is *computable* if the infinite binary sequence $\mathrm{Binary}(\alpha)$ is computable. For example, every rational is a computable real, and the reals $\sqrt{2}$ and π are computable as well. We can then show that, for every real α, the following conditions are equivalent:

(i) The real α is computable.
(ii) The infinite binary sequence $\mathrm{Binary}_R(\alpha)$ is computable.
(iii) The infinite binary sequence $\mathrm{Binary}_L(\alpha)$ is computable.
(iv) There exists a total recursive function $f : \mathbb{N}^+ \to \mathbb{Q}$ such that $|\alpha - f(n)| < 1/n$ for all $n \in \mathbb{N}^+$.[2]
(v) There exists a total recursive function $f : \mathbb{N}^+ \to \mathbb{Z}$ such that $|\alpha - f(n)/n| < 1/n$ for all $n \in \mathbb{N}^+$.
(vi) The real α is both left-computable and right-computable.
(vii) The mapping $\mathbb{N} \ni n \mapsto \lfloor \alpha n \rfloor$ is a total recursive function.
(viii) The mapping $\mathbb{N} \ni n \mapsto \lceil \alpha n \rceil$ is a total recursive function.

Since every rational is computable, it follows that if a real α is not computable then $\mathrm{Binary}_R(\alpha) = \mathrm{Binary}_L(\alpha)$.

---

[1] This condition is just a paraphrase of the definition of the recursive enumerability of a real α using the term "total recursive function."

[2] This condition can be equivalently rephrased as the condition that there exists a computable sequence $\{a_n\}_{n \in \mathbb{N}}$ of rationals such that $|\alpha - a_n| < 2^{-n}$ for all $n \in \mathbb{N}$.

A sequence $\{a_n\}_{n\in\mathbb{N}}$ of reals is called *computable* if there exists a total recursive function $f: \mathbb{N} \times \mathbb{N} \to \mathbb{Q}$ such that $|a_n - f(n, m)| < 2^{-m}$ for all $n, m \in \mathbb{N}$.

See Pour-El and Richards [28] for the detail of the treatment of the computability of reals and sequences of reals.

2.3 Prefix-Free Machines and Program-Size Complexity

In this section we review some definitions and results of algorithmic information theory (AIT, for short). As a prerequisite for reading this book, we refer the reader to Chaitin [9] for a treatment of the development of AIT.

A *prefix-free machine* is a partial recursive function $M: \{0, 1\}^* \to \{0, 1\}^*$ such that dom M is a prefix-free set. For each prefix-free machine M and each $s \in \{0, 1\}^*$, we define $K_M(s)$ by the following.[3]

$$K_M(s) := \min\{|p| \mid p \in \{0, 1\}^* \ \& \ M(p) = s\}.$$

A prefix-free machine U is said to be *optimal* if for each prefix-free machine M there exists $d \in \mathbb{N}$, which depends on M, with the following property; for every $p \in$ dom M there exists $q \in \{0, 1\}^*$ for which $U(q) = M(p)$ and $|q| \leq |p| + d$. We can then show that there exists an optimal prefix-free machine (see Theorem 2.2 of Chaitin [9] for the proof). We choose an arbitrary particular optimal prefix-free machine U *as the standard one for use throughout the rest of the book.* We then define $K(s)$ as $K_U(s)$, which is referred to as the *program-size complexity* of s, the *information content* of s, or the *Kolmogorov complexity* of s (Gács [16], Levin [21], Chaitin [9]). It follows that for every prefix-free machine M there exists $d \in \mathbb{N}$ such that, for every $s \in \{0, 1\}^*$,

$$K(s) \leq K_M(s) + d. \tag{2.4}$$

We call an element of dom U a *program* for U.

Let s be an arbitrary finite binary string. Then, consider a prefix-free machine M_0 such that dom $M_0 = \{\lambda\}$ and $M_0(\lambda) = s$. Since U is optimal, it follows from the definition of the optimality that there exists $p \in \{0, 1\}^*$ for which $U(p) = M_0(\lambda)$. Thus, for every $s \in \{0, 1\}^*$ there is $p \in$ dom U such that $U(p) = s$. Based on this fact, we see that $\lambda \notin$ dom U, in particular. This is because, if $\lambda \in$ dom U, then dom $U = \{\lambda\}$ since dom U is prefix-free. Hence, we have that $K(s) \in \mathbb{N}^+$ for every $s \in \{0, 1\}^*$.

For any $n \in \mathbb{N}$, $K(n)$ is defined to be K(the nth element of $\{0, 1\}^*$). We choose a particular bijective total recursive function $b: \{0, 1\}^* \times \{0, 1\}^* \to \{0, 1\}^*$ *as the standard one for use throughout the rest of the book.* We then define $K(s, t)$ as $K(b(s, t))$ for any $s, t \in \{0, 1\}^*$.

Consider a prefix-free machine M_1 such that, for every $p, s \in \{0, 1\}^*$, $M_1(p) = s$ if and only if $p = qs$ and $U(q) = |s|$ for some $q \in$ dom U. Then, applying (2.4) to

---

[3]In the case where there does not exist $p \in$ dom M with $M(p) = s$, we set $K_M(s) := \infty$.

M_1, we have that there exists $c \in \mathbb{N}$ such that, for every $s \in \{0, 1\}^*$,

$$K(s) \le |s| + K(|s|) + c. \tag{2.5}$$

Here, the term $K(|s|)$ comes from the fact that dom U is a prefix-free set.

Next, consider a prefix-free machine M_2 such that, for every $p, s \in \{0, 1\}^*$, $M_2(p) = s$ if and only if $|s| \ge 1$ and $p = 0s_1 0s_2 \cdots 0s_{k-1} 1 s_k$ where $s = s_1 s_2 \cdots s_{k-1} s_k$ with each $s_i \in \{0, 1\}$. Then, applying (2.4) to M_2, we have that $K(s) \le 2|s| + O(1)$ for all $s \in \{0, 1\}^*$ with $|s| \ge 1$. Therefore, there exists $c \in \mathbb{N}$ such that, for every $n \in \mathbb{N}^+$,

$$K(n) \le 2\log_2 n + c. \tag{2.6}$$

Thus, it follows from (2.5) that there exists $d \in \mathbb{N}$ such that

$$K(s) \le |s| + 2\log_2 |s| + d \tag{2.7}$$

for every $s \ne \lambda$ (Li and Vitányi [22]).

Let $\Psi \colon \{0, 1\}^* \to \{0, 1\}^*$ be a partial recursive function. Consider a prefix-free machine M_3 such that, for every $p, t \in \{0, 1\}^*$, $M_3(p) = t$ if and only if $\Psi(U(p)) = t$. Then, applying (2.4) to M_3, we have that there exists $c \in \mathbb{N}$ such that, for every $s \in$ dom Ψ,

$$K(\Psi(s)) \le K(s) + c. \tag{2.8}$$

The program-size complexity $K(s)$ is originally defined using the concept of program-size, as stated above. However, it is possible to define $K(s)$ without referring to such a concept. To be specific, as in the following, we first introduce a *universal probability* m, and then define $K(s)$ as $-\log_2 m(s)$. A universal probability is defined as follows.

Definition 2.1 (*Universal Probability, Zvonkin and Levin* [58]) A real-valued function $r \colon \{0, 1\}^* \to [0, 1]$ is called a *lower-computable semi-measure* if the following conditions (i) and (ii) hold:

(i) $\sum_{s \in \{0,1\}^*} r(s) \le 1$.
(ii) The set $\{(a, s) \in \mathbb{Q} \times \{0, 1\}^* \mid a < r(s)\}$ is r.e., or equivalently, there exists a total recursive function $f \colon \mathbb{N}^+ \times \{0, 1\}^* \to \mathbb{Q}$ such that, for each $s \in \{0, 1\}^*$, it holds that $\lim_{n \to \infty} f(n, s) = r(s)$ and $\forall n \in \mathbb{N}^+ \ f(n, s) \le r(s)$.

We say that a lower-computable semi-measure m is a *universal probability* if for every lower-computable semi-measure r, there exists $c \in \mathbb{N}^+$ such that, for all $s \in \{0, 1\}^*$, it holds that $r(s) \le cm(s)$. □

The following theorem can be then shown (see Theorem 3.4 of Chaitin [9] for its proof). For each prefix-free machine M and $s \in \{0, 1\}^*$, we define $P_M(s)$ as $\sum_{M(p)=s} 2^{-|p|}$.

Theorem 2.1 *Let V be an optimal prefix-free machine. Then both $2^{-K_V(s)}$ and $P_V(s)$ are universal probabilities as a function of $s \in \{0, 1\}^*$.* □

By Theorem 2.1, we see that

$$K(s) = -\log_2 m(s) + O(1) \tag{2.9}$$

for every universal probability m. Thus it is possible to define $K(s)$ as $-\log_2 m(s)$ with a particular universal probability m instead of as $K_U(s)$. Note that the difference up to an additive constant is nonessential to AIT.

Chaitin [9] showed Theorems 2.2 and 2.3 below, together known as the *Kraft-Chaitin Theorem*.

Theorem 2.2 (Kraft-Chaitin Theorem I, Chaitin [9]) *Let $f : \mathbb{N} \to \mathbb{N}$ be a total recursive function such that $\sum_{n=0}^{\infty} 2^{-f(n)} \leq 1$. Then there exists a total recursive function $g : \mathbb{N} \to \{0, 1\}^*$ such that (i) g is an injection, (ii) the set $\{g(n) \mid n \in \mathbb{N}\}$ is prefix-free, and (iii) $|g(n)| = f(n)$ for all $n \in \mathbb{N}$.* □

For the proof of Theorem 2.2, see the proof of Theorem 3.2 of Chaitin [9]. Theorem 2.2 results in Theorem 2.3 as follows.

Theorem 2.3 (Kraft-Chaitin Theorem II, Chaitin [9]) *Let both $f : \mathbb{N} \to \mathbb{N}$ and $g : \mathbb{N} \to \{0, 1\}^*$ be total recursive functions. Suppose that $\sum_{n=0}^{\infty} 2^{-f(n)} \leq 1$. Then there exists a prefix-free machine M which satisfies the following two conditions:*

(i) *For every $l \in \mathbb{N}$ and $s \in \{0, 1\}^*$, it holds that*

$$\#\{p \in \{0, 1\}^* \mid |p| = l \ \& \ M(p) = s\} = \#\{n \in \mathbb{N} \mid f(n) = l \ \& \ g(n) = s\}.$$

(ii) $K_M(s) = \min\{f(n) \mid g(n) = s\}$ *for all $s \in g(\mathbb{N})$.*

Proof First, it follows from Theorem 2.3 that there exists a total recursive function $h : \mathbb{N} \to \{0, 1\}^*$ such that h is an injection, the set $\{h(n) \mid n \in \mathbb{N}\}$ is prefix-free, and $|h(n)| = f(n)$ for all $n \in \mathbb{N}$. Consider a prefix-free machine M such that, for every $p, s \in \{0, 1\}^*$, $M(p) = s$ if and only if there exists $n \in \mathbb{N}$ such that $p = h(n)$ and $s = g(n)$. Note that such a prefix-free machine M exists due to the properties of the total recursive function h.

Since h is an injection, for each $l \in \mathbb{N}$ and $s \in \{0, 1\}^*$, we see that

$$\#\{p \in \{0, 1\}^* \mid |p| = l \ \& \ M(p) = s\} = \#\{n \in \mathbb{N} \mid |h(n)| = l \ \& \ s = g(n)\}.$$

Thus, since $|h(n)| = f(n)$ for all $n \in \mathbb{N}$, we have the condition (i) for M in Theorem 2.3. The condition (ii) for M in Theorem 2.3 follows from the condition (i) for M in Theorem 2.3. □

2.4 Equivalence of the Notions of Randomness for Reals

As stated in Sect. 2.1, we identify a real with an infinite binary sequence. Specifically, we identify a real α with the infinite binary sequence Binary (α). The notion of program-size complexity plays an important role in characterizing the randomness of an infinite binary sequence, i.e., a real, as follows.

Definition 2.2 (*Weak Chaitin Randomness, Chaitin* [9, 12]) For any $\omega \in \{0, 1\}^\infty$, we say that ω is *weakly Chaitin random* if there exists $c \in \mathbb{N}$ such that

$$n - c \leq K(\omega\lceil_n)$$

for all $n \in \mathbb{N}^+$. For any $\alpha \in \mathbb{R}$, we say that α is *weakly Chaitin random* if Binary (α) is weakly Chaitin random.   □

Since the finite binary string $\omega\lceil_n$ has the length of n, in the above definition we regard an infinite binary sequence ω as *random* if an arbitrary prefix $\omega\lceil_n$ of ω cannot be compressed by the program-size complexity $K(s)$ *in a substantial way*, i.e., *up to an additive constant*.

Thus, in Definition 2.2 we have defined the notion of randomness for an infinite binary sequence based the incompressibility of it by means of program-size complexity. However, it is possible to define the notion of randomness for an infinite binary sequence in a completely different manner with no reference to the notion of program-size complexity. We can do this based on the notion of an *effective null set*. An effective null set is a set of infinite binary sequences which has a Lebesgue measure of zero and moreover has a certain computational property. In the alternative definition of the randomness, we regard an infinite binary sequence as *random* if it does not belong to any effective null set. We can define several types of the notions of randomness for an infinite binary sequence, depending on what computational property we impose on a null set in the definition of effective null set.

Among all randomness notions based on effective null sets, *Martin-Löf randomness* is a central one nowadays, and is the oldest historically. In particular, the Martin-Löf randomness is equivalent to the weak Chaitin randomness, as stated in Theorem 2.4 below. The notion of Martin-Löf randomness is defined as follows, based on the notion of *Martin-Löf test*, which is a specific implementation of the notion of effective null set.

Definition 2.3 (*Martin-Löf Randomness, Martin-Löf* [24]) A *Martin-Löf test* is a subset $\mathscr{C}$ of $\mathbb{N}^+ \times \{0, 1\}^*$ such that (i) $\mathscr{C}$ is an r.e. set and (ii) for every $n \in \mathbb{N}^+$ it holds that

$$\sum_{s \in \mathscr{C}_n} 2^{-|s|} \leq 2^{-n},\tag{2.10}$$

where $\mathscr{C}_n$ denotes the set $\left\{ s \mid (n, s) \in \mathscr{C} \right\}$.

For any $\omega \in \{0, 1\}^\infty$, we say that ω is *Martin-Löf random* if for every Martin-Löf test $\mathscr{C}$, there exists $n \in \mathbb{N}^+$ such that, for every $k \in \mathbb{N}^+$, it holds that $\omega\lceil_k \notin \mathscr{C}_n$.

For any $\alpha \in \mathbb{R}$, we say that α is *Martin-Löf random* if Binary (α) is Martin-Löf random. $\square$

Let $\mathscr{C}$ be a Martin-Löf test. It follows from the condition (2.10) that the set $\bigcap_{n=1}^{\infty} [\mathscr{C}_n]^{\prec}$ forms a null set. The recursive enumerability of $\mathscr{C}$ gives a computational property to this set in the definition. In this manner, the set $\bigcap_{n=1}^{\infty} [\mathscr{C}_n]^{\prec}$ plays a role as an effective null set in Definition 2.3.

From the definition of Martin-Löf randomness, we can show, for example, that every Martin-Löf random infinite binary sequence is Borel normal, by means of an appropriate Martin-Löf test.

The following theorem is one of the main results of AIT, which states that the two definitions of the randomness for an infinite binary sequence introduced from different places are equivalent to each other.

Theorem 2.4 (Schnorr [33], Chaitin [9]) *For every* $\omega \in \{0, 1\}^{\infty}$, ω *is weakly Chaitin random if and only if* ω *is Martin-Löf random.* $\square$

The equivalence supports our intuition that each of the weak Chaitin randomness and the Martin-Löf randomness is appropriate as the definition of the randomness of an infinite binary sequence. In the next chapter, we provide a generalization of Theorem 2.4 over the notion of *partial randomness*.

In 1975 Chaitin [9] introduced an concrete example of a random real, denoted Ω. His Ω number is defined as follows.

$$\Omega_V := \sum_{p \in \text{dom } V} 2^{-|p|}$$

for each optimal prefix-free machine V. Since dom V is prefix-free, Ω_V converges and satisfies $0 < \Omega_V \leq 1$ due to the Kraft inequality (2.1). In particular, based on our specific optimal prefix-free machine U, we define Ω as Ω_U. The real Ω is called *Chaitin's* Ω, or *Chaitin's halting probability* Ω.

The real Ω is interpreted as the *halting probability* of U, i.e., the probability that the optimal prefix-free machine U halts when U starts on the input tape filled with an infinite binary sequence generated by infinitely repeated tosses of a fair coin. The detail is as follows: The class of prefix-free machines is equal to the class of functions which are computed by *self-delimiting Turing machines*. A self-delimiting Turing machine is a deterministic Turing machine which has two tapes, a program tape and a work tape. The program tape is infinite to the right, while the work tape is infinite in both directions. An input string in $\{0, 1\}^*$ is put on the program tape. See Chaitin [9] for the detail of self-delimiting Turing machines. Let $\mathscr{U}$ be a self-delimiting Turing machine which computes the optimal prefix-free machine U. Then, the universal probability $P_U(s)$, introduced in the previous section, is the probability that $\mathscr{U}$ halts and outputs s when $\mathscr{U}$ starts on the program tape filled with an infinite binary sequence generated by infinitely repeated tosses of a fair coin. Therefore, since $\Omega = \sum_{s \in \{0,1\}^*} P_U(s)$ holds, the real Ω is *the probability that* $\mathscr{U}$ *just halts* under the same setting.

From the definition of Ω we can show that, given $\Omega\!\restriction_n$, one can effectively calculate all programs p for U with $|p| \le n$. By discovering this property of Ω, Chaitin proved the following theorem.

Theorem 2.5 (Chaitin [9]) *The real Ω is weakly Chaitin random.*   □

Thus, by Theorem 2.4, Chaitin's Ω is Martin-Löf random. In the next chapter, we provide a generalization of Theorem 2.5 over the notion of *partial randomness*.

Based on the notion of program-size complexity, we can define another notion of the randomness for an infinite binary sequence, which at first glance appears stronger than the notion of weak Chaitin randomness, as in the following form.

Definition 2.4 (*Chaitin Randomness, Chaitin* [9, 12]) For any $\omega \in \{0, 1\}^\infty$, we say that ω is *Chaitin random* if $\lim_{n \to \infty} K(\omega\!\restriction_n) - n = \infty$. For any $\alpha \in \mathbb{R}$, we say that α is *Chaitin random* if Binary (α) is Chaitin random.   □

Actually, we can show that, for every $\alpha \in \mathbb{R}$, α is weakly Chaitin random if and only if α is Chaitin random (see Chaitin [12] for the proof and historical detail). Thus Ω is Chaitin random.

Chapter 3
Partial Randomness

3.1 *D*-Randomness

At the turn of the 21st century, Tadaki [40, 41] generalized Chaitin's halting probability Ω to $\Omega(D)$ as follows.

$$\Omega(D) := \sum_{p \in \text{dom } U} 2^{-|p|/D}, \tag{3.1}$$

where D is an arbitrary positive real. Thus, $\Omega(1) = \Omega$ obviously. If $0 < D \leq 1$, then $\Omega(D)$ converges and $0 < \Omega(D) < 1$ since $\Omega(D) \leq \Omega < 1$.

To see the randomness property of $\Omega(D)$, let us introduce first the notion of the compression rate of a real.

Definition 3.1 Let ω be an infinite binary sequence. If the limit value

$$\lim_{n \to \infty} \frac{K(\omega\!\restriction_n)}{n}$$

exists, we call it the *compression rate* of ω. The *compression rate* of a real α is defined as the compression rate of Binary (α). □

The idea behind the definition is as follows: Note that n is the length of $\omega\!\restriction_n$ before the compression by means of program-size complexity while $K(\omega\!\restriction_n)$ is the length of $\omega\!\restriction_n$ at the time of the compression. Therefore, the ratio $K(\omega\!\restriction_n)/n$ is considered to be the compression rate of the finite binary string $\omega\!\restriction_n$. Hence, on letting $n \to \infty$, the limit value of the ratio is considered to be the compression rate of the infinite binary sequence ω.

Using the inequality (2.7), we can see that the compression rate of an infinite binary sequence lies between 0 and 1 certainly, if it exists. From Definition 2.2, we

This chapter is based on Tadaki [40, 41, 44, 45], and is a rearrangement of the parts of them.

© The Author(s), under exclusive license to Springer Nature Singapore Pte Ltd. 2019
K. Tadaki, *A Statistical Mechanical Interpretation of Algorithmic Information Theory*,
SpringerBriefs in Mathematical Physics 36,
https://doi.org/10.1007/978-981-15-0739-7_3

see that every weakly Chaitin random infinite binary sequence has the maximum compression rate of 1. Thus, the compression rate of Chaitin's Ω equals to 1. On the other hand, every computable infinite binary sequence has the minimum compression rate of 0. To see this, let ω be an arbitrary computable infinite binary sequence. Then, consider a total recursive function $\Psi \colon \mathbb{N}^+ \to \{0, 1\}^*$ with $\Psi(n) = \omega{\upharpoonright}_n$. Using the inequalities (2.8) and (2.7), we see that

$$K(\omega{\upharpoonright}_n) = K(\Psi(n)) \le K(n) + O(1) \le \log_2 n + 2\log_2 \log_2 n + O(1)$$

for all $n \in \mathbb{N}^+$. Thus, we have that every computable infinite binary sequence has the compression rate of 0.

A concrete real which has an intermediate compression rate between 0 and 1 is $\Omega(D)$ introduced above. Specifically, Theorem 3.1 below holds for $\Omega(D)$. At this point, let us introduce the notion of *partial randomness* in order to treat the notion of the compression rate *more finely*. Actually, Theorem 3.1 is stated based on the notion of the partial randomness. Tadaki [40, 41] generalized the notion of the randomness of an infinite binary sequence so that *the degree of the randomness*, which is referred to as *the partial randomness* nowadays [5, 6, 14, 29], can be characterized by a real D with $0 \le D \le 1$. In the case of $D = 1$, the notion of partial randomness results in the notion of randomness for an infinite binary sequence introduced in the preceding section. The notion of partial randomness is used in most theorems in the rest of the book, as well as in Theorem 3.1. Compared with the notion of the compression rate, the notion of partial randomness is natural and easy to treat, as we see in what follows.

Definition 3.2 (*Weak Chaitin D-Randomness, Tadaki* [40, 41]) Let D be a real with $0 \le D \le 1$. For any $\omega \in \{0, 1\}^\infty$, we say that ω is *weakly Chaitin D-random* if there exists $c \in \mathbb{N}$ such that, for all $n \in \mathbb{N}^+$, it holds that $Dn - c \le K(\omega{\upharpoonright}_n)$. For any $\alpha \in \mathbb{R}$, we say that α is *weakly Chaitin D-random* if Binary (α) is weakly Chaitin D-random. $\square$

Definition 3.3 (*Chaitin D-Randomness, Tadaki* [40, 41]) Let D be a real with $0 \le D \le 1$. For any $\omega \in \{0, 1\}^\infty$, we say that ω is *Chaitin D-random* if it holds that $\lim_{n \to \infty} K(\omega{\upharpoonright}_n) - Dn = \infty$. For any $\alpha \in \mathbb{R}$, we say that α is *Chaitin D-random* if Binary (α) is Chaitin D-random. $\square$

Obviously, for every $\omega \in \{0, 1\}^\infty$, if ω is Chaitin D-random then ω is weakly Chaitin D-random. However, Reimann and Stephan [29] showed that, in the case of $D < 1$, the converse does not necessarily hold. On the other hand, in the case where $D = 1$, the weak Chaitin 1-randomness and the Chaitin 1-randomness obviously result in the weak Chaitin randomness and the Chaitin randomness, respectively. In this case, we can show that an infinite binary sequence is weakly Chaitin 1-random if and only if it is Chaitin 1-random, as stated at the end of the preceding chapter.

Definition 3.4 (*D-compressibility, Tadaki* [40, 41]) Let D be a real with $0 \le D \le 1$. For any $\omega \in \{0, 1\}^\infty$, we say that ω is *D-compressible* if $K(\omega{\upharpoonright}_n) \le Dn + o(n)$ for all $n \in \mathbb{N}^+$, i.e., if it holds that

$$\limsup_{n\to\infty} \frac{K(\omega\restriction_n)}{n} \leq D.$$

For any $\alpha \in \mathbb{R}$, we say that α is *D-compressible* if Binary (α) is *D*-compressible. $\qquad\square$

Let ω be an arbitrary infinite binary sequence. If ω is weakly Chaitin *D*-random and *D*-compressible then the compression rate of ω is *D*, obviously. However, the condition that ω has the compression rate of *D* does not necessarily imply that ω is weakly Chaitin *D*-random. In this sense, the notion of partial randomness is a stronger representation of the notion of compression rate.

Theorem 3.1 (Tadaki [40, 41])

   (i) *If* $0 < D \leq 1$ *and D is computable, then* $\Omega(D)$ *converges and is weakly Chaitin D-random and D-compressible. Therefore, the compression rate of* $\Omega(D)$ *equals to D in this case.*
   (ii) *If* $1 < D$, *then* $\Omega(D)$ *diverges to* ∞. $\qquad\square$

Thus, $\Omega(D)$ has the following property: As *D* becomes larger while satisfying $0 < D \leq 1$, the partial randomness of $\Omega(D)$ increases. In the case when $D = 1$, the real $\Omega(D)$ results in Ω and becomes the most random. Since there does not exist a real whose compression rate is greater than one, $\Omega(D)$ has to diverge if $D > 1$.

Proof (of Theorem 3.1) (i) Suppose that *D* is a computable real with $0 < D \leq 1$. Let $p_1, p_2, p_3, \ldots$ be a recursive enumeration of the r.e. set dom *U*. We denote $\Omega(D)$ by α.

We first show that $\Omega(D)$ is weakly Chaitin *D*-random. The proof is obtained by generalizing Chaitin's original proof that Ω is weakly Chaitin random. Since *D* is a computable real, it follows from (2.3) that there exists a partial recursive function $\xi : X \to \mathbb{N}^+$ such that, for every $n \in \mathbb{N}^+$, it holds that

$$0.\left(\alpha\restriction_n^L\right) < \sum_{i=1}^{\xi\left(\alpha\restriction_n^L\right)} 2^{-|p_i|/D}.$$

Using (2.3) again, it is then easy to see that, for every $n, i \in \mathbb{N}^+$, if $i > \xi\left(\alpha\restriction_n^L\right)$ then $Dn < |p_i|$. In other words, given $\alpha\restriction_n^L$, one can calculate all programs *p* for *U* of length at most $\lfloor Dn \rfloor$. Therefore, for every $n \in \mathbb{N}^+$ and $s \in \{0, 1\}^*$, if $s \neq U(p_i)$ for all $i \leq \xi\left(\alpha\restriction_n^L\right)$ then $Dn < K(s)$. Hence, given $\alpha\restriction_n^L$, by calculating the set $\left\{ U(p_i) \mid i \leq \xi\left(\alpha\restriction_n^L\right) \right\}$ and picking any particular finite binary string that is not in this set, one can obtain an $s \in \{0, 1\}^*$ such that $Dn < K(s)$.

Thus, there exists a partial recursive function $\Psi : \{0, 1\}^* \to \{0, 1\}^*$ such that, for every $n \in \mathbb{N}^+$, it holds that $Dn < K(\Psi(\alpha\restriction_n^L))$. Using (2.8), there is $c \in \mathbb{N}$ such that, for every $n \in \mathbb{N}^+$, $K(\Psi(\alpha\restriction_n^L)) \leq K(\alpha\restriction_n^L) + c$. Therefore, $\text{Binary}_L(\alpha)$ is weakly Chaitin *D*-random. It follows that $\text{Binary}_L(\alpha)$ is not computable, which implies that $\text{Binary}_L(\alpha) = \text{Binary}(\alpha)$. Thus, $\Omega(D)$ is weakly Chaitin *D*-random.

Next, we prove that $\Omega(D)$ is D-compressible. Since D is computable, we note that there exists a total recursive function $f : \mathbb{N}^+ \times \mathbb{N} \to \mathbb{N}$ such that, for every $k \in \mathbb{N}^+$ and $n \in \mathbb{N}$, it holds that

$$\left| \sum_{i=1}^{k} 2^{-|p_i|/D} - 2^{-n} f(k,n) \right| < 2^{-n}. \tag{3.2}$$

Due to the result above, $\Omega(1)$ (i.e., Ω) is weakly Chaitin random. We denote $\Omega(1)$ by β.

Given n and $\beta\!\restriction_{\lceil Dn \rceil}$ one can find a k_0 with the property that

$$0. \left(\beta\!\restriction_{\lceil Dn \rceil} \right) < \sum_{i=1}^{k_0} 2^{-|p_i|}.$$

It is then easy to see that

$$\sum_{i=k_0+1}^{\infty} 2^{-|p_i|} < 2^{-Dn}.$$

Using the inequality $x^z + y^z \le (x+y)^z$ for reals $x, y > 0$ and $z \ge 1$, it follows that

$$\left| \Omega(D) - \sum_{i=1}^{k_0} 2^{-|p_i|/D} \right| < 2^{-n}. \tag{3.3}$$

From (3.2), (3.3), and $|\Omega(D) - 0.(\alpha\!\restriction_n)| < 2^{-n}$ it is shown that

$$\left| 0.(\alpha\!\restriction_n) - 2^{-n} f(k_0, n) \right| < 3 \cdot 2^{-n}.$$

Hence $\alpha\!\restriction_n = f(k_0, n), f(k_0, n) \pm 1, f(k_0, n) \pm 2$, where $\alpha\!\restriction_n$ is regarded as a dyadic integer. Based on this, one is left with five possibilities of $\alpha\!\restriction_n$, so that one needs only 3 bits more in order to determine $\alpha\!\restriction_n$.

Thus, there exists a partial recursive function $\Phi : \mathbb{N}^+ \times \{0,1\}^* \times \{0,1\}^* \to \{0,1\}^*$ such that for every $n \in \mathbb{N}^+$ there exists $s \in \{0,1\}^3$ with the property that $\Phi(n, \beta\!\restriction_{\lceil Dn \rceil}, s) = \alpha\!\restriction_n$. Consider a prefix-free machine M such that, for every $p, v \in \{0,1\}^*$, $M(p) = v$ if and only if there exist $q \in \text{dom } U, t \in \{0,1\}^*, s \in \{0,1\}^3$, and $n \in \mathbb{N}^+$ with the properties that $p = qts$, $U(q) = n$, $|t| = \lceil Dn \rceil$, and $\Phi(n, t, s) = v$. Note that such a prefix-free machine M exists since D is computable. Then, it is easy to see that

$$K_M(\alpha\!\restriction_n) \le K(n) + |\beta\!\restriction_{\lceil Dn \rceil}| + 3$$

for every $n \in \mathbb{N}^+$. Thus, since $\lim_{n \to \infty} K(n)/n = 0$ due to (2.7), and also $|\beta\!\restriction_{\lceil Dn \rceil}| \le Dn + 1$ holds, it follows from (2.4) that $\Omega(D)$ is D-compressible.

(ii) Suppose that $D > 1$. We then choose a particular computable real d satisfying $D \geq d > 1$. Contrarily, assume that $\Omega(d)$ converges. Based on an argument similar to the first half of the proof of Theorem 3.1 (i), we see that $\Omega(d)$ is weakly Chaitin d-random, i.e., there exists $c \in \mathbb{N}$ such that $dn - c \leq K(\Omega(d)\!\restriction_n)$. It follows from (2.7) that $dn - c \leq n + o(n)$. Dividing by n and letting $n \to \infty$ we have $d \leq 1$, which contradicts the fact $d > 1$. Thus, $\Omega(d)$ diverges to infinity. By noting $\Omega(d) \leq \Omega(D)$, we have that $\Omega(D)$ diverges to infinity. $\qquad\square$

As we saw above, the number $\Omega(D)$ is an instance of a real which is *weakly* Chaitin D-random and D-compressible, for any computable real D with $0 < D \leq 1$. In the next section, we give an instance of a real which is Chaitin D-random and D-compressible, for any computable real D with $0 < D < 1$.

The notion of Martin-Löf randomness can also be generalized over the notion of partial randomness as follows.

Definition 3.5 (*Martin-Löf D-Randomness, Tadaki* [41]) Let D be a real with $0 \leq D \leq 1$. A *Martin-Löf D-test* is a subset $\mathscr{C}$ of $\mathbb{N}^+ \times \{0, 1\}^*$ such that (i) $\mathscr{C}$ is an r.e. set and (ii) for every $n \in \mathbb{N}^+$ it holds that

$$\sum_{s \in \mathscr{C}_n} 2^{-D|s|} \leq 2^{-n},$$

where $\mathscr{C}_n$ denotes the set $\big\{ s \mid (n, s) \in \mathscr{C} \big\}$.

For any $\omega \in \{0, 1\}^\infty$, we say that ω is *Martin-Löf D-random* if for every Martin-Löf D-test $\mathscr{C}$, there exists $n \in \mathbb{N}^+$ such that, for every $k \in \mathbb{N}^+$, it holds that $\omega\!\restriction_k \notin \mathscr{C}_n$. For any $\alpha \in \mathbb{R}$, we say that α is *Martin-Löf D-random* if Binary (α) is Martin-Löf D-random. $\qquad\square$

Thus, in the case where $D = 1$, the Martin-Löf D-randomness results in the Martin-Löf randomness given in Definition 2.3.

We can generalize Theorem 2.4 over the notion of D-randomness as follows.

Theorem 3.2 (Tadaki [41]) *Let D be a computable real with $0 \leq D \leq 1$. Then, for every $\omega \in \{0, 1\}^\infty$, ω is weakly Chaitin D-random if and only if ω is Martin-Löf D-random.* $\qquad\square$

In the case where $D = 1$, Theorem 3.2 results in Theorem 2.4, which is Theorem R1 in Chaitin [12]. The proof of Theorem 3.2 is obtained by generalizing the proof of Theorem R1, given in Chaitin [12]. Thus, in order to prove Theorem 3.2, we need Theorem 3.3 below and Theorem 2.3, both of which were shown in Chaitin [9].

Theorem 3.3 (Chaitin [9]) *There is $c \in \mathbb{N}$ such that, for every $n \in \mathbb{N}$ and $k \in \mathbb{Z}$, it holds that $\#\{ s \in \{0, 1\}^* \mid |s| = n \ \& \ K(s) < k \} \leq 2^{k - K(n) + c}$.* $\qquad\square$

Theorem 3.3 above is Theorem 4.2 (b) of Chaitin [9]. The proof of Theorem 3.2 is then given as follows.

Proof (of Theorem 3.2) Suppose that D is a computable real and $D \geq 0$. Let $f \colon \mathbb{N} \to \mathbb{N}$ with $f(n) = \lfloor Dn \rfloor$. Then f is a total recursive function.

First, we show that the Martin-Löf D-randomness implies the weak Chaitin D-randomness. Actually, we show the contraposition. Suppose that ω is not weak Chaitin D-random. Then, for every $k \in \mathbb{N}$ there is $n \in \mathbb{N}^+$ such that $K(\omega\!\restriction_n) < f(|\omega\!\restriction_n|) - k$. Let $\mathscr{T} = \{(k, s) \in \mathbb{N} \times \{0, 1\}^* \mid K(s) < f(|s|) - k - c\}$ for the natural number c which is referred to in Theorem 3.3. Then, for every $k \in \mathbb{N}$ there exists $n \in \mathbb{N}^+$ such that $\omega\!\restriction_n \in \mathscr{T}_k$, where $\mathscr{T}_k$ denotes the set $\{s \mid (k, s) \in \mathscr{T}\}$. On the other hand, it follows from Theorem 3.3 that $\#\{s \in \mathscr{T}_k \mid |s| = n\} \leq 2^{Dn - K(n) - k}$ for every $k, n \in \mathbb{N}$. Hence, for each $k \in \mathbb{N}^+$ we get

$$\sum_{s \in \mathscr{T}_k} 2^{-D|s|} = \sum_{n=0}^{\infty} \#\{s \in \mathscr{T}_k \mid |s| = n\} 2^{-Dn} \leq 2^{-k} \left(\sum_{n=0}^{\infty} 2^{-K(n)}\right) \leq 2^{-k} \Omega < 2^{-k}.$$

Since f is a total recursive function, $\mathscr{T}$ is an r.e. set. Thus, $\mathscr{T}$ is Martin-Löf D-test. Hence, ω is not Martin-Löf D-random, as desired.

Next, we show that the weak Chaitin D-randomness implies the Martin-Löf D-randomness. Actually, we show the contraposition, again. Thus, we suppose that ω is not Martin-Löf D-random. Then, there exists a Martin-Löf D-test $\mathscr{T}$ such that, for every $n \in \mathbb{N}^+$, there exists $k \in \mathbb{N}^+$ for which $\omega\!\restriction_k \in \mathscr{T}_n$. We then see that

$$\sum_{n=1}^{\infty} \sum_{s \in \mathscr{T}_{2n+1}} 2^{-[f(|s|) - n]} \leq \sum_{n=1}^{\infty} \left(2^{n+1} \sum_{s \in \mathscr{T}_{2n+1}} 2^{-D|s|}\right) \leq \sum_{n=1}^{\infty} 2^{-n} = 1.$$

Since $\mathscr{T}$ is an r.e. set, there exists an injective total recursive function $g \colon \mathbb{N} \to \mathbb{N} \times \{0, 1\}^*$ such that $g(\mathbb{N}) = \{(n, s) \mid n \in \mathbb{N}^+ \ \& \ s \in \mathscr{T}_{2n+1}\}$. Let $g_1 \colon \mathbb{N} \to \mathbb{N}$ and $g_2 \colon \mathbb{N} \to \{0, 1\}^*$ be total recursive functions such that $g(k) = (g_1(k), g_2(k))$ for all $k \in \mathbb{N}$. Then

$$\sum_{k=0}^{\infty} 2^{-[f(|g_2(k)|) - g_1(k)]} = \sum_{n=1}^{\infty} \sum_{s \in \mathscr{T}_{2n+1}} 2^{-[f(|s|) - n]} \leq 1.$$

Since f is a total recursive function, it follows from Theorem 2.3 that there is a prefix-free machine M such that $K_M(s) = \min\{f(|g_2(k)|) - g_1(k) \mid g_2(k) = s\}$ for all $s \in g_2(\mathbb{N})$. Therefore, using (2.4), we have that there exists $d \in \mathbb{N}$ such that, for every $n \in \mathbb{N}^+$ and every $s \in \mathscr{T}_{2n+1}$, it holds that $K(s) \leq D|s| - n + d$. Thus, since for every $n \in \mathbb{N}^+$ there exists $k \in \mathbb{N}^+$ such that $\omega\!\restriction_k \in \mathscr{T}_{2n+1}$, we see that for every $n \in \mathbb{N}^+$ there exists $k \in \mathbb{N}^+$ such that $K(\omega\!\restriction_k) \leq D|\omega\!\restriction_k| - n + d = Dk - n + d$. Hence, ω is not weakly Chaitin D-random, as desired. $\qquad\square$

3.2 Chaitin D-Randomness and Divergence

For any reals $Q > 0$ and $D > 0$, we define $W(Q, D)$ by

$$W(Q, D) := \sum_{p \in \text{dom } U} |p|^Q 2^{-|p|/D}. \tag{3.4}$$

We can show the following theorem as to the notion of *Chantin D-randomness*, instead of *weak Chantin D-randomness* as in Theorem 3.1.

Theorem 3.4 (Tadaki [44]) *Let Q and D be positive reals.*

(i) *If $0 < D < 1$ and both Q and D are computable, then $W(Q, D)$ converges and is a left-computable real which is Chaitin D-random and D-compressible.*
(ii) *If $1 \leq D$, then $W(Q, D)$ diverges to ∞.* $\qquad\square$

The techniques used in the proofs of Theorems 3.1 and 3.4 are frequently used throughout the rest of the book as basic tools. We see that the weak Chaitin D-randomness in Theorem 3.1 is replaced by the Chaitin D-randomness in Theorem 3.4 in exchange for the divergence at $D = 1$. In order to derive this divergence we make use of Theorem 3.5 (i) below. We prove Theorem 3.5 in a general form. For example, using Theorem 3.5, we can show that the *Shannon entropy*

$$- \sum_{s \in \{0,1\}^*} m(s) \log_2 m(s) \tag{3.5}$$

of an arbitrary universal probability m diverges to ∞.

We say that a function $f \colon \mathbb{N}^+ \to [0, \infty)$ is *lower-computable* if there exists a total recursive function $a \colon \mathbb{N}^+ \times \mathbb{N}^+ \to \mathbb{Q}$ such that, for each $n \in \mathbb{N}^+$, it holds that $\lim_{k \to \infty} a(k, n) = f(n)$ and $\forall k \in \mathbb{N}^+ \ a(k, n) \leq f(n)$.

Theorem 3.5 (Tadaki [44, 45]) *Let A be an infinite r.e. subset of $\{0, 1\}^*$ and let $f \colon \mathbb{N}^+ \to [0, \infty)$ be a lower-computable function such that $\lim_{n \to \infty} f(n) = \infty$. Then the following hold.*

(i) *$\sum_{U(p) \in A} f(|p|) 2^{-|p|}$ diverges to ∞, where the sum is over all programs p for U such that $U(p) \in A$.*
(ii) *If there exists $l_0 \in \mathbb{N}^+$ such that $f(l) 2^{-l}$ is a non-increasing function of l for all $l \geq l_0$, then $\sum_{s \in A} f(K(s)) 2^{-K(s)}$ diverges to ∞.*

Proof (i) Contrarily, assume that $\sum_{U(p) \in A} f(|p|) 2^{-|p|}$ converges. Then, there exists $d \in \mathbb{N}^+$ such that $\sum_{U(p) \in A} f(|p|) 2^{-|p|} \leq d$. We define a function $r \colon \{0, 1\}^* \to [0, \infty)$ by

$$r(s) = \frac{1}{d} \sum_{U(p) = s} f(|p|) 2^{-|p|}$$

if $s \in A$; $r(s) = 0$ otherwise. Then we see that $\sum_{s \in \{0,1\}^*} r(s) \leq 1$ and therefore r is a lower-computable semi-measure. Since $P_U(s)$ is a universal probability by Theorem 2.1, there exists $c \in \mathbb{N}^+$ such that $r(s) \leq c P_U(s)$ for all $s \in \{0,1\}^*$. Hence we have

$$\sum_{U(p)=s} (cd - f(|p|))2^{-|p|} \geq 0 \tag{3.6}$$

for all $s \in A$. On the other hand, since A is an infinite set and $\lim_{n \to \infty} f(n) = \infty$, there is $s_0 \in A$ such that $f(|p|) > cd$ for all p with $U(p) = s_0$. Therefore we have $\sum_{U(p)=s_0} (cd - f(|p|))2^{-|p|} < 0$. However, this contradicts (3.6), and the proof of Theorem 3.5 (i) is completed.

(ii) We first note that there is $n_0 \in \mathbb{N}$ such that $K(s) \geq l_0$ for all s with $|s| \geq n_0$. Now, let us assume contrarily that $\sum_{s \in A} f(K(s))2^{-K(s)}$ converges. Then, there exists $d \in \mathbb{N}^+$ such that $\sum_{s \in A} f(K(s))2^{-K(s)} \leq d$. We define a function $r \colon \{0,1\}^* \to [0, \infty)$ by

$$r(s) = \frac{1}{d} f(K(s))2^{-K(s)}$$

if $s \in A$ and $|s| \geq n_0$; $r(s) = 0$ otherwise. Then we see that $\sum_{s \in \{0,1\}^*} r(s) \leq 1$ and therefore r is a lower-computable semi-measure. Since $2^{-K(s)}$ is a universal probability by Theorem 2.1, there exists $c \in \mathbb{N}^+$ such that $r(s) \leq c2^{-K(s)}$ for all $s \in \{0,1\}^*$. Hence, if $s \in A$ and $|s| \geq n_0$, then $cd \geq f(K(s))$. On the other hand, since A is an infinite set and $\lim_{n \to \infty} f(n) = \infty$, there is $s_0 \in A$ such that $|s_0| \geq n_0$ and $f(K(s_0)) > cd$. Thus, we have a contradiction, and the proof of Theorem 3.5 (ii) is completed. $\square$

Corollary 3.1 (Tadaki [44, 45]) *If m is a universal probability and A is an infinite r.e. subset of $\{0,1\}^*$, then $-\sum_{s \in A} m(s) \log_2 m(s)$ diverges to ∞.*

Proof We first note that there is a real $x_0 > 0$ such that the function $x2^{-x}$ of a real x is decreasing for $x \geq x_0$. For this x_0, there is $n_0 \in \mathbb{N}$ such that $-\log_2 m(s) \geq x_0$ for all s with $|s| \geq n_0$. On the other hand, by (2.9), there is $c \in \mathbb{N}$ such that $-\log_2 m(s) \leq K(s) + c$ for all $s \in \{0,1\}^*$. Thus, we see that

$$- \sum_{s \in A \& |s| \geq n_0} m(s) \log_2 m(s) \geq \sum_{s \in A \& |s| \geq n_0} (K(s) + c)2^{-K(s)-c}$$

$$= 2^{-c} \sum_{s \in A \& |s| \geq n_0} K(s)2^{-K(s)} + c2^{-c} \sum_{s \in A \& |s| \geq n_0} 2^{-K(s)}.$$

It follows from Theorem 3.5 (ii) that $\sum_{s \in A} K(s)2^{-K(s)}$ diverges to ∞. Hence, we see, by the inequality above, that $-\sum_{s \in A} m(s) \log_2 m(s)$ also diverges to ∞. $\square$

By Corollary 3.1, we see that the Shannon entropy (3.5) of an arbitrary universal probability m diverges to ∞.

Now, the proof of Theorem 3.4 is given as follows.

Proof (of Theorem 3.4) Let Q and D be positive reals, and let $p_1, p_2, p_3, \ldots$ be an enumeration of the infinite set dom U. We then have $W(Q, D) = \lim_{m \to \infty} W_m$ (Q, D), where

$$W_m(Q, D) := \sum_{i=1}^{m} |p_i|^Q 2^{-|p_i|/D}.$$

Note that the enumeration $\{p_i\}$ can be chosen quite arbitrarily. This is because the infinite sum (3.4) is a positive term series, and therefore the limit value $W(Q, D)$ is independent of the choice of $\{p_i\}$ regardless of whether it converges or not. Thus, in what follows, we choose $\{p_i\}$ to be a recursive enumeration of dom U, in particular, for the sake of convenience.

(i) Suppose that Q and D are computable and $D < 1$.

First we show that $W(Q, D)$ converges to a left-computable real. Since $D < 1$, there is $l_0 \in \mathbb{N}^+$ such that

$$\frac{1}{D} - \frac{Q \log_2 l}{l} \geq 1$$

for all $l \geq l_0$. Then there is $m_0 \in \mathbb{N}^+$ such that $|p_i| \geq l_0$ for all $i > m_0$. Thus, we see that, for each $i > m_0$,

$$|p_i|^Q 2^{-|p_i|/D} = 2^{-(\frac{1}{D} - \frac{Q \log_2 |p_i|}{|p_i|})|p_i|} \leq 2^{-|p_i|}.$$

Hence, for each $m > m_0$,

$$W_m(Q, D) - W_{m_0}(Q, D) = \sum_{i=m_0+1}^{m} |p_i|^Q 2^{-|p_i|/D} \leq \sum_{i=m_0}^{m} 2^{-|p_i|} < \Omega.$$

Thus, since $\{W_m(Q, D)\}_m$ is an increasing sequence of reals, it converges to a real $W(Q, D)$ as $m \to \infty$. Moreover, since Q and D are computable, $W(Q, D)$ is shown to be left-computable.

We then show that $W(Q, D)$ is Chaitin D-random. We denote $W(Q, D)$ by α. It follows from (2.3) that $\lceil W(Q, D) \rceil - 1 + 0.\left(\alpha|_n^L\right) < W(Q, D)$ for all $n \in \mathbb{N}^+$. Since Q and D are computable reals, there exists a partial recursive function $\xi \colon \{0, 1\}^* \to \mathbb{N}^+$ such that, for every $n \in \mathbb{N}^+$, it holds that

$$\lceil W(Q, D) \rceil - 1 + 0.\left(\alpha|_n^L\right) < W_{\xi(\alpha|_n^L)}(Q, D).$$

It is then easy to see that

$$W(Q, D) - W_{\xi(\alpha|_n^L)}(Q, D) < 2^{-n}$$

for every $n \in \mathbb{N}^+$. Therefore, for each $n, i \in \mathbb{N}^+$, we see that if $i > \xi(\alpha|_n^L)$ then

$$|p_i|^Q 2^{-|p_i|/D} < 2^{-n}$$

and therefore $QD \log_2 |p_i| < |p_i| - Dn$. Thus, given $\alpha\lceil_n^L$, by calculating the set $\{ U(p_i) \mid i \leq \xi(\alpha\lceil_n^L) \}$ and picking any particular finite binary string which is not in this set, one can obtain an $s \in \{0, 1\}^*$ such that $QD \log_2 K(s) < K(s) - Dn$.

Hence, there exists a partial recursive function $\Psi : \{0, 1\}^* \to \{0, 1\}^*$ such that, for every $n \in \mathbb{N}^+$, it holds that $QD \log_2 K(\Psi(\alpha\lceil_n^L)) < K(\Psi(\alpha\lceil_n^L)) - Dn$. Applying this inequality to itself, we have $QD \log_2 n < K(\Psi(\alpha\lceil_n^L)) - Dn + O(1)$. On the other hand, it follows from (2.8) that there is $c \in \mathbb{N}$ such that, for every $n \in \mathbb{N}^+$, it holds that $K(\Psi(\alpha\lceil_n^L)) < K(\alpha\lceil_n^L) + c$. Therefore, we have

$$QD \log_2 n < K(\alpha\lceil_n^L) - Dn + O(1).$$

Hence, $\mathrm{Binary}_L(\alpha)$ is Chaitin D-random. It follows that $\mathrm{Binary}_L(\alpha)$ is not computable, which implies that $\mathrm{Binary}_L(\alpha) = \mathrm{Binary}(\alpha)$. Thus, $W(Q, D)$ is Chaitin D-random.

Next, we show that $W(Q, D)$ is D-compressible. Since Q and D are computable reals, there exists a total recursive function $g : \mathbb{N}^+ \times \mathbb{N}^+ \to \mathbb{Z}$ such that, for every $m, n \in \mathbb{N}^+$, it holds that

$$\left| W_m(Q, D) - \lfloor W(Q, D) \rfloor - 2^{-n} g(m, n) \right| < 2^{-n}. \tag{3.7}$$

Let d be an arbitrary computable real with $D < d < 1$. Then, due to the results above, the limit value $W(Q, d)$ exists as a real and is Chaitin d-random. We denote $W(Q, d)$ by β.

Given n and $\beta\lceil_{\lceil Dn/d \rceil}$, one can find an $m_0 \in \mathbb{N}^+$ such that

$$\lceil W(Q, d) \rceil - 1 + 0.(\beta\lceil_{\lceil Dn/d \rceil}) < W_{m_0}(Q, d).$$

This is possible since $\lceil W(Q, d) \rceil - 1 + 0.(\beta\lceil_{\lceil Dn/d \rceil}) < W(Q, d)$ due to (2.3) and $\lim_{m \to \infty} W_m(Q, d) = W(Q, d)$ holds. It is then easy to see that

$$\sum_{i=m_0+1}^{\infty} |p_i|^Q 2^{-|p_i|/d} < 2^{-Dn/d}.$$

Raising both sides of this inequality to the power d/D and using the inequality $x^z + y^z \leq (x + y)^z$ for reals $x, y > 0$ and $z \geq 1$, we see that

$$2^{-n} > \sum_{i=m_0+1}^{\infty} |p_i|^{Qd/D} 2^{-|p_i|/D} > \sum_{i=m_0+1}^{\infty} |p_i|^Q 2^{-|p_i|/D}.$$

It follows that

$$\left| W(Q, D) - W_{m_0}(Q, D) \right| < 2^{-n}. \tag{3.8}$$

From (3.7), (3.8), and $|\lfloor W(Q, D) \rfloor + 0.(W(Q, D)\lceil_n) - W(Q, D)| < 2^{-n}$, it is shown that $| W(Q, D)\lceil_n - g(m_0, n) | < 3$. Therefore,

$$W(Q, D)\upharpoonright_n = g(m_0, n),\ g(m_0, n) \pm 1,\ g(m_0, n) \pm 2$$

where $W(Q, D)\upharpoonright_n$ is regarded as a dyadic integer. Thus, one is left with five possibilities of $W(Q, D)\upharpoonright_n$, so that one needs only 3 bits more in order to determine $W(Q, D)\upharpoonright_n$.

Thus, there exists a partial recursive function $\Phi \colon \mathbb{N}^+ \times \{0, 1\}^* \times \{0, 1\}^* \to \{0, 1\}^*$ such that for every $n \in \mathbb{N}^+$ there exists $s \in \{0, 1\}^3$ with the property that $\Phi(n, \beta\upharpoonright_{\lceil Dn/d \rceil}, s) = W(Q, D)\upharpoonright_n$. Consider a prefix-free machine M such that, for every $p, v \in \{0, 1\}^*$, $M(p) = v$ if and only if there exist $q \in \mathrm{dom}\, U$, $t \in \{0, 1\}^*$, $s \in \{0, 1\}^3$, and $n \in \mathbb{N}^+$ with the properties that $p = qts$, $U(q) = n$, $|t| = \lceil Dn/d \rceil$, and $\Phi(n, t, s) = v$. Note that such a prefix-free machine M exists since D and d are computable. Then, it is easy to see that

$$K_M(W(Q, D)\upharpoonright_n) \leq K(n) + |\beta\upharpoonright_{\lceil Dn/d \rceil}| + 3$$

for every $n \in \mathbb{N}^+$. Thus, since $\lim_{n \to \infty} K(n)/n = 0$ due to (2.7), and also $|\beta\upharpoonright_{\lceil Dn/d \rceil}| \leq Dn/d + 1$ holds, it follows from (2.4) that $W(Q, D)$ is D/d-compressible. Since d is an arbitrary computable real with $D < d < 1$, it follows that $W(Q, D)$ is D-compressible.

(ii) We choose any particular computable real Q' with $Q \geq Q' > 0$. Then, using Theorem 3.5 (i), we can show that $W(Q', 1) = \sum_{p \in \mathrm{dom}\, U} |p|^{Q'} 2^{-|p|}$ diverges to ∞. Suppose that $D \geq 1$. Then, since $W_m(Q, D) \geq W_m(Q', 1)$ for every $m \in \mathbb{N}^+$, we see that $W(Q, D)$ diverges to ∞. $\qquad \square$

3.3 Algorithmic Dimension

The notion of *Hausdorff dimension* plays a central role in fractal geometry (see e.g. Falconer [15]). The notions of the compression rate by means of program-size complexity and partial randomness which we have studied so far are known to be equivalent to Hausdorff dimension. The equivalence of them was first revealed by the pioneering work [31, 32] of Ryabko in the 1980s. Subsequently, the study of the equivalence was developed by the work [38, 39] of Staiger in the 1990s. Since 2000, the study of the equivalence has been further developed through the works of Tadaki [40, 41], Lutz [23], and Mayordomo [25].

In this process of development of the subject, through the works [40, 41] we introduced six new fractal dimensions, called *algorithmic dimensions*, based on the partial randomness or the compression rate by means of program-size complexity. Each of the six algorithmic dimensions is a fractal dimension defined for a subset F of the N-dimensional Euclidean space $\mathbb{R}^N$. Among them, the strongest one is the notion of *the first algorithmic dimension*, which is defined based on the notion of partial randomness as follows.

Definition 3.6 (*Tadaki* [40, 41]) Let F be a subset of $\mathbb{R}^N$. The *first algorithmic dimension* $\dim_{A1} F$ of F is defined as a real D such that

(i) the infinite binary sequence x is Chaitin $\frac{D}{N}$-random for some $x \in F$, and

(ii) the infinite binary sequence x is $\frac{D}{N}$-compressible for all $x \in F$.　　□

In the definition, an element x of F is regarded as an infinite binary sequence, according to the identification of $\mathbb{R}^N$ with $\{0, 1\}^\infty$ in the following manner (Tadaki [40, 41]): First, we denote a point x of $\mathbb{R}^N$ by its components as $x = (x^1, x^2, \ldots, x^N)$, and then expand the fractional part of each x^i in base two as follows.

$$x^i - \lfloor x^i \rfloor = 0.x_1^i x_2^i x_3^i x_4^i \ldots\ldots\ldots .$$

We then enumerate each bit x_k^i of the base-two expansions in the following order.

$$x_1^1 x_1^2 \ldots x_1^N x_2^1 x_2^2 \ldots x_2^N x_3^1 x_3^2 \ldots x_3^N x_4^1 x_4^2 \ldots x_4^N \ldots\ldots\ldots . \tag{3.9}$$

Finally, we identify the point x of $\mathbb{R}^N$ with this infinite binary sequence.

Among the six algorithmic dimensions, two of the weakest ones are the notions of *upper algorithmic dimension* and *lower algorithmic dimension*, which are defined as follows.

Definition 3.7 (*Tadaki* [41]) Let F be a subset of $\mathbb{R}^N$. The *upper algorithmic dimension* $\overline{\dim}_A F$ of F and the *lower algorithmic dimension* $\underline{\dim}_A F$ of F are, respectively, defined by

$$\overline{\dim}_A F := \sup_{x \in F} \limsup_{n \to \infty} \frac{K(x\lceil_n)}{n/N},$$

$$\underline{\dim}_A F := \sup_{x \in F} \liminf_{n \to \infty} \frac{K(x\lceil_n)}{n/N}.$$
　　　　　　　　　　　　　　　　　　　　　　　　　　　　　　　　□

In the definition, $x\lceil_n$ denotes a generalization of $\alpha\lceil_n$, which is originally defined for a real α in Sect. 2.1, over a point x of $\mathbb{R}^N$; that is, $x\lceil_n$ is defined as the first n bits of the infinite binary sequence (3.9), according to our identification of $\mathbb{R}^N$ with $\{0, 1\}^\infty$ explained above.

Tadaki [40, 41] showed that all the six algorithmic dimensions are equal to the Hausdorff dimension for any *self-similar set* which is *computable* in a certain sense. The class of such self-similar sets includes familiar fractal sets such as the Cantor set, von Koch curve, and Sierpiński gasket. In this manner, the notion of partial randomness (and therefore the compression rate by means of program-size complexity) is equivalent to the notion of Hausdorff dimension. The study of this equivalence forms one of the main subjects in the field of AIT nowadays.

In light of the results of the book, this equivalence between partial randomness and dimension is *further extended* to include the notion of *temperature*. Actually, in the subsequent chapters, we will see that the notion of temperature in statistical mechanics is equivalent to partial randomness. Thus, the three elements; *partial randomness, dimension,* and *temperature* are closely related to one another as *mutually equivalent notions*.

Chapter 4
Temperature Equals to Partial Randomness

4.1 Superficial Similarity Between $\Omega(D)$ and Partition Function

In 2006 Calude and Stay [6] pointed out a *superficial* similarity of $\Omega(D)$ to a partition function in statistical mechanics. In statistical mechanics, the *partition function* $Z_{sm}(T)$ at temperature T is defined by

$$Z_{sm}(T) := \sum_{n \in \mathcal{E}} \exp\left\{-\frac{E_n}{k_B T}\right\}, \tag{4.1}$$

where $\mathcal{E}$ is a complete set of energy eigenstates of a quantum system and E_n is the energy of an energy eigenstate n. Recall that the partition function $Z_{sm}(T)$ is of particular importance in equilibrium statistical mechanics. This is because all the thermodynamic quantities of the system can be expressed by using the partition function $Z_{sm}(T)$, and the knowledge of $Z_{sm}(T)$ is sufficient to understand all the macroscopic properties of the system (Toda et al. [57], Callen [2]).

Calude and Stay [6] pointed out, in essence, that the partition function $Z_{sm}(T)$ has *superficially* the same form as $\Omega(D)$ by performing the following replacements in $Z_{sm}(T)$.

Replacements 4.1

(i) *Replace the complete set $\mathcal{E}$ of energy eigenstates n by the set* dom U *of all programs p for U.*

(ii) *Replace the energy E_n of an energy eigenstate n by the length $|p|$ of a program p.*

This chapter is based on Tadaki [44], and is a rearrangement of the parts of it.

© The Author(s), under exclusive license to Springer Nature Singapore Pte Ltd. 2019
K. Tadaki, *A Statistical Mechanical Interpretation of Algorithmic Information Theory*,
SpringerBriefs in Mathematical Physics 36,
https://doi.org/10.1007/978-981-15-0739-7_4

(iii) Set the Boltzmann constant k_B to $1/\ln 2$, where the $\ln$ denotes the natural logarithm.   □

By performing Replacements 4.1, the partition function $Z_{sm}(T)$ is converted into the form

$$\sum_{p \in \text{dom } U} 2^{-|p|/T}.$$

If we further identify the temperature T as the partial randomness D, we have $\Omega(D)$ certainly. The factor $\exp\{-E_n/(k_B T)\}$ which appears in the definition (4.1) of the partition function is called the *Boltzmann factor*. Note that the Boltzmann factor $\exp\{-E_n/(k_B T)\}$ is converted into $2^{-|p|/T}$ by Replacements 4.1.

In this book, inspired by this superficial similarity between $\Omega(D)$ in AIT and a partition function in statistical mechanics, we develop a *statistical mechanical interpretation* of AIT, where $\Omega(D)$ appears as a partition function. For that purpose, first we review the basic framework of equilibrium statistical mechanics further, following Sect. 1.3. In particular, we review the notion of canonical ensemble and thermodynamic quantities.

4.2 Canonical Ensemble and Thermodynamic Quantities

In this section, we review the notion of thermodynamic quantities in statistical mechanics. We introduce them, based on the argument developed in Sect. 1.3. Note that the argument in this section is on the same level of mathematical strictness as statistical mechanics in physics, as in Chap. 1. Thus, we ignore mathematical rigor in this section.

In quantum mechanics, any quantum system is described by a quantum state completely. In statistical mechanics, among all quantum states, energy eigenstates are of particular importance. If a quantum system is in an energy eigenstate, then the quantum system has a definite energy. Then, the fundamental postulate of statistical mechanics is the *principle of equal probability*, which is stated in Sect. 1.3.

As we saw in Sect. 1.3, the notion of temperature is introduced into a quantum system in the following manner: First, the *entropy* $S_{sm}(E)$ of a quantum system with energy E is defined by

$$S_{sm}(E) := k_B \ln \Theta(E). \tag{4.2}$$

Here, $\Theta(E)$ is the total number of energy eigenstates of the quantum system whose energy E' satisfies that $E \le E' \le E + \delta E$. Then, the *temperature* T of the quantum system with energy E is defined by

$$\frac{1}{T} := \frac{\partial S_{sm}}{\partial E}(E).$$

Now, let us consider a quantum system Q at constant temperature T. To be precise, consider a quantum system Q in thermal contact with an another very large quantum system Q_R, called a *heat reservoir*, whose temperature is T. By applying the principle of equal probability to the composite quantum system consisting of the two quantum systems Q and Q_R we have the following consequence.

Consequence of the Principle of Equal Probability: The probability $\Pr(n)$ that the quantum system Q is in an energy eigenstate n with energy E_n is given by

$$\Pr(n) = \frac{1}{Z_{sm}(T)} \exp\left\{-\frac{E_n}{k_B T}\right\},$$

where the partition function $Z_{sm}(T)$, defined by (4.1), plays a role as a normalization factor. The distribution $\Pr(n)$ is called the *canonical ensemble* or the *canonical distribution*. □

Based on the consequence above, the *thermodynamic quantities* of the quantum system Q at temperature T are calculated as follows. First, the *energy* $E_{sm}(T)$ of the quantum system Q at temperature T is the expected value of the energy E_n of an energy eigenstate n of the quantum system Q when the quantum system Q is realized in an energy eigenstate n according to the canonical distribution $\Pr(n)$. Thus, we have

$$E_{sm}(T) = \sum_{n \in \mathscr{E}} E_n \Pr(n) = \frac{1}{Z_{sm}(T)} \sum_{n \in \mathscr{E}} E_n \exp\left\{-\frac{E_n}{k_B T}\right\}. \qquad (4.3)$$

Next, based on the above consequence, we can show that the *Helmholtz free energy* $F_{sm}(T)$ of the quantum system Q at temperature T has the form

$$F_{sm}(T) = -k_B T \ln Z_{sm}(T). \qquad (4.4)$$

Here, the Helmholtz free energy is a thermodynamic quantity which equals to the maximum amount of work that the quantum system Q can perform against the external system during a process at constant temperature T. Then, based on the above consequence, we can show that the *entropy* $S_{sm}(T)$ of the quantum system Q at temperature T is given by

$$S_{sm}(T) = \frac{E_{sm}(T) - F_{sm}(T)}{T}. \qquad (4.5)$$

Note here that the entropy $S_{sm}(T)$ equals to the *Shannon entropy* of the probability distribution $\{\Pr(n)\}$. To be precise, it is easy to show that

$$S_{sm}(T) = -k_B \sum_{n \in \mathscr{E}} \Pr(n) \ln \Pr(n).$$

Finally, the *specific heat* $C_{sm}(T)$ of the quantum system Q at temperature T is given by

$$C_{sm}(T) = \frac{d}{dT} E_{sm}(T). \tag{4.6}$$

4.3 Thermodynamic Quantities of AIT

The purpose of this book is to study what happens when we perform Replacements 4.1 for the thermodynamic quantities $E_{sm}(T)$, $F_{sm}(T)$, $S_{sm}(T)$, and $C_{sm}(T)$ of a quantum system at temperature T presented in the preceding section. Although the argument in the preceding section is on the same level of mathematical strictness as statistical mechanics in physics, we make an argument in a fully mathematically rigorous manner *throughout the rest of the book, except for Chap.* 6.

We choose an arbitrary particular enumeration $q_1, q_2, q_3, q_4, \ldots$ of the countably infinite set dom U *as the standard one for use throughout the rest of this chapter.*[1] Then, motivated by the formulae (4.3), (4.4), (4.5), and (4.6) and taking into account Replacements 4.1, we introduce the notion of thermodynamic quantities into AIT as follows. Here we redefine $\Omega(D)$ as the partition function $Z(T)$ of AIT.[2]

Definition 4.1 (*Thermodynamic Quantities of AIT, Tadaki* [44]) Let T be any real with $T > 0$.

(i) The *partition function* $Z(T)$ of AIT at temperature T is defined as $\lim_{k \to \infty} Z_k(T)$ where

$$Z_k(T) := \sum_{i=1}^{k} 2^{-|q_i|/T}.$$

(ii) The *energy* $E(T)$ of AIT at temperature T is defined as $\lim_{k \to \infty} E_k(T)$ where

$$E_k(T) := \frac{1}{Z_k(T)} \sum_{i=1}^{k} |q_i| \, 2^{-|q_i|/T}.$$

(iii) The *Helmholtz free energy* $F(T)$ of AIT at temperature T is defined as $\lim_{k \to \infty} F_k(T)$ where

$$F_k(T) := -T \log_2 Z_k(T).$$

(iv) The *entropy* $S(T)$ of AIT at temperature T is defined as $\lim_{k \to \infty} S_k(T)$ where

$$S_k(T) := \frac{E_k(T) - F_k(T)}{T}.$$

---

[1] The enumeration $\{q_i\}$ is quite arbitrary and therefore we do not, ever, require $\{q_i\}$ to be a recursive enumeration of dom U.

[2] To be precise, the partition function is not a thermodynamic quantity but a statistical mechanical quantity.

(v) The *specific heat* $C(T)$ of AIT at temperature T is defined as $\lim_{k \to \infty} C_k(T)$ where $C_k(T) := E_k'(T)$, the derived function of $E_k(T)$. □

At this point of time, let us consider the meaning of (iii) of Replacements 4.1. When we perform (iii) of Replacements 4.1 for the entropy $S_{sm}(E)$ defined by (4.2), it is converted into the form

$$S_{sm}(E) = \log_2 \Theta(E). \tag{4.7}$$

On the other hand, recall that in information theory (Shannon [34], Ash [1], Cover and Thomas [13]), the (Shannon) entropy is expressed in bits, and therefore the log is to the base 2 in its definition, just as in the Eq. (4.7) above. Thus, (iii) of Replacements 4.1 corresponds to this fact in information theory. AIT has a close relationship to information theory as indicated by the name (see Chaitin [9] for the detail). Thus, (iii) of Replacements 4.1 is considered to reflect this close relationship.

We investigate the convergence and partial randomness of the thermodynamic quantities of AIT. By elaborating the proofs of Theorems 3.1 and 3.4, we can show the following two theorems.

Theorem 4.2 (Tadaki [40, 41, 44]) *Let T be a real with $T > 0$.*

(i) *If $0 < T \le 1$ and T is computable, then each of $Z(T)$ and $F(T)$ converges and is weakly Chaitin T-random and T-compressible.*

(ii) *If $1 < T$, then $Z(T)$ and $F(T)$ diverge to ∞ and $-\infty$, respectively.* □

Theorem 4.3 (Tadaki [44]) *Let T be a real with $T > 0$.*

(i) *If $0 < T < 1$ and T is computable, then each of $E(T)$, $S(T)$, and $C(T)$ converges and is Chaitin T-random and T-compressible.*

(ii) *If $T = 1$, then all of $E(T)$, $S(T)$, and $C(T)$ diverge to ∞.*

(iii) *If $1 < T$, then both $E(T)$ and $S(T)$ diverge to ∞.*[3] □

The statements about $Z(T)$ in Theorem 4.2 is precisely Theorem 3.1. The statements about $F(T)$ in Theorem 4.2, and all the statements in Theorem 4.3 are proved in the next section.

Theorems 4.2 and 4.3 show that if T is a computable real with $0 < T < 1$, then the *partial randomness* (and therefore the *compression rate*) of the values of all the thermodynamic quantities in Definition 4.1 equals to the *temperature T*. These theorems also show that the values of all the thermodynamic quantities diverge when the temperature T exceeds 1. This phenomenon might be regarded as a certain sort of *phase transition* in statistical mechanics. Comparing Theorem 4.3 with Theorem 4.2, we see that the weak Chaitin T-randomness in (i) of Theorem 4.2 is strengthened to the Chaitin T-randomness in (i) of Theorem 4.3. However, checking the behavior at temperature $T = 1$, we see that while $Z(1)$ and $F(1)$ converge in Theorem 4.2, the quantities $E(1)$, $S(1)$, and $C(1)$ diverge to ∞ in Theorem 4.3 in compensation for the strengthen.

---

[3] It is still open whether $C(T)$ diverges or not in the case of $T > 1$.

In Chap. 8 we introduce the notion of the *strict T*-compressibility for a real (see Definition 8.5), which is a strengthening of the notion of T-compressibility. Actually, the T-compressibility of $Z(T)$ and $F(T)$ in Theorem 4.2 can be strengthened to the strict T-compressibility in the case where $0 < T < 1$ and T is computable. This will be proved in Chap. 8, based on Theorem 7.5 proved in Chap. 7. On the other hand, note that the T-compressibility of $E(T)$, $S(T)$, and $C(T)$ in Theorem 4.3 cannot be strengthened to the strict T-compressibility in the case where $0 < T < 1$ and T is computable, since any Chaitin T-random real cannot be strict T-compressible.

4.4 The Proofs of Theorems 4.2 and 4.3

In what follows, we prove the statements about $F(T)$ in Theorem 4.2, and all the statements in Theorem 4.3. Before that, we remark that the following theorem holds for $Z(T)$, obviously.

Theorem 4.4 (Tadaki [40, 41]) *If $0 < T \le 1$ and T is computable, then $Z(T)$ converges and is a left-computable real.* □

4.4.1 Helmholtz Free Energy

We prove the following theorem, which leads to the statements about $F(T)$ in Theorem 4.2.

Theorem 4.5 (Tadaki [44]) *Let $T \in \mathbb{R}$.*

  (i) *If $0 < T \le 1$ and T is computable, then $F(T)$ converges and is a positive right-computable real which is weakly Chaitin T-random and T-compressible.*
 (ii) *If $1 < T$, then $F(T)$ diverges to $-\infty$.*

Proof (i) Suppose that T is a computable real with $0 < T \le 1$.

Since $Z(T)$ converges and $Z(T) > 0$, it follows from Definition 4.1 that $F(T)$ also converges and satisfies that

$$F(T) = -T \log_2 Z(T). \tag{4.8}$$

By this equality we see that the real $F(T)$ is independent of the choice of the enumeration of $\{q_i\}$ of dom U in Definition 4.1.

The real $Z(T)$ is left-computable due to Theorem 4.4. Thus, since T is a right-computable real and $-\log_2 Z(T) > 0$, it follows that $F(T)$ is a positive right-computable real.

We then show that $F(T)$ is weakly Chaitin T-random. Using the mean value theorem and (4.8), it is easy to see that there exists $c \in \mathbb{N}$ such that, for every $A, B \in \mathbb{R}$, if $A \ge F(T)$ and $B \ge 1/T$ then

$$0 \leq Z(T) - 2^{-AB} \leq 2^c \max\{A - F(T), B - 1/T\}. \tag{4.9}$$

Since $F(T)$ is a right-computable real, there exists a total recursive function $f: \mathbb{N}^+ \to \mathbb{Q}$ such that $F(T) \leq f(m)$ for all $m \in \mathbb{N}^+$ and $\lim_{m \to \infty} f(m) = F(T)$. Since T is a computable real, there exists a total recursive function $g: \mathbb{N}^+ \to \mathbb{Q}$ such that $0 \leq g(n) - 1/T < 2^{-n}$ for all $n \in \mathbb{N}^+$.

Given $F(T) \upharpoonright_n$, one can find an $m_0 \in \mathbb{N}^+$ such that

$$f(m_0) < \lfloor F(T) \rfloor + 0.(F(T) \upharpoonright_n) + 2^{-n}.$$

This is possible due to (2.2). It follows that $0 \leq f(m_0) - F(T) < 2^{-n}$. Therefore, by (4.9) it is shown that $0 \leq Z(T) - 2^{-f(m_0)g(n)} < 2^{c-n}$. One can then calculate the finite binary string $2^{-f(m_0)g(n)} \upharpoonright_n$. This is possible since $2^{-f(m_0)g(n)}$ is a real of the form 2^q with $q \in \mathbb{Q}$. It follows from (2.2) that

$$0 \leq 0.(Z(T) \upharpoonright_n) - 0.l_n < Z(T) - 2^{-f(m_0)g(n)} + 2^{-n} < (2^c + 1)2^{-n},$$

where $2^{-f(m_0)g(n)} \upharpoonright_n$ is denoted by l_n. Hence $Z(T) \upharpoonright_n = l_n, l_n + 1, l_n + 2, \ldots, l_n + 2^c$, where $Z(T) \upharpoonright_n$ and l_n are regarded as a dyadic integer. Thus, one is left with $2^c + 1$ possibilities of $Z(T) \upharpoonright_n$, so that one needs only $c + 1$ bits more in order to determine $Z(T) \upharpoonright_n$.

Thus, there exists a partial recursive function $\Phi: \{0, 1\}^* \times \{0, 1\}^* \to \{0, 1\}^*$ such that for every $n \in \mathbb{N}^+$ there exists $s \in \{0, 1\}^*$ with the properties that $|s| = c + 1$ and $\Phi(F(T) \upharpoonright_n, s) = Z(T) \upharpoonright_n$. Consider a prefix-free machine M such that, for every $p, v \in \{0, 1\}^*$, $M(p) = v$ if and only if there exist $q \in \mathrm{dom}\, U$, $s \in \{0, 1\}^{c+1}$, and $t \in \{0, 1\}^*$ with the properties that $p = qs$, $U(q) = t$, and $\Phi(t, s) = v$. Note that such a prefix-free machine M exists. Then, it is easy to see that

$$K_M(Z(T) \upharpoonright_n) \leq K(F(T) \upharpoonright_n) + c + 1$$

for every $n \in \mathbb{N}^+$. It follows from (2.4) that

$$K(Z(T) \upharpoonright_n) \leq K(F(T) \upharpoonright_n) + O(1). \tag{4.10}$$

Thus, since $Z(T)$ is weakly Chaitin T-random by Theorem 3.1 (i), $F(T)$ is also weakly Chaitin T-random.

Next, we show that $F(T)$ is T-compressible. Since T is a computable real, there exists a total recursive function $a: \{0, 1\}^* \times \mathbb{N}^+ \to \mathbb{Z}$ such that, for every $s \in \{0, 1\}^*$ and $n \in \mathbb{N}^+$, if $0.s > 0$ then

$$\left| -T \log_2 0.s - \lfloor F(T) \rfloor - 2^{-n} a(s, n) \right| < 2^{-n}. \tag{4.11}$$

By the mean value theorem, it is also shown that there is $k \in \mathbb{N}^+$ such that, for every $n \in \mathbb{N}^+$, if $0.(Z(T) \upharpoonright_n) > 0$ then

$$\left| -T \log_2 0.(Z(T)\upharpoonright_n) - F(T) \right| < 2^{k-n}. \qquad (4.12)$$

From (4.11), (4.12), and $|\lfloor F(T) \rfloor + 0.(F(T)\upharpoonright_n) - F(T)| < 2^{-n}$, it is shown that, for every $n \in \mathbb{N}^+$, if $0.(Z(T)\upharpoonright_n) > 0$ then $|F(T)\upharpoonright_n - a(Z(T)\upharpoonright_n, n)| < 2^k + 2$ and therefore

$$F(T)\upharpoonright_n = a(Z(T)\upharpoonright_n, n), \ a(Z(T)\upharpoonright_n, n) \pm 1, \ a(Z(T)\upharpoonright_n, n) \pm 2, \dots,$$
$$a(Z(T)\upharpoonright_n, n) \pm (2^k + 1),$$

where $F(T)\upharpoonright_n$ is regarded as a dyadic integer. Hence, given $Z(T)\upharpoonright_n$, one is left with $2^{k+1} + 3$ possibilities of $F(T)\upharpoonright_n$ to determine it, so that one needs only $k + 2$ bits more in order to determine $F(T)\upharpoonright_n$.

Thus, there exists a partial recursive function $\Psi : \{0, 1\}^* \times \{0, 1\}^* \rightarrow \{0, 1\}^*$ such that for every $n \in \mathbb{N}^+$ there exists $s \in \{0, 1\}^*$ with the properties that $|s| = k + 2$ and $\Psi(Z(T)\upharpoonright_n, s) = F(T)\upharpoonright_n$. Based on this, we can show that

$$K(F(T)\upharpoonright_n) \le K(Z(T)\upharpoonright_n) + O(1),$$

in a similar manner used in deriving the inequality (4.10) above. Thus, since $Z(T)$ is T-compressible by Theorem 3.1 (i), $F(T)$ is also T-compressible.

(ii) In the case of $T > 1$, since $\lim_{k \to \infty} Z_k(T) = \infty$ by Theorem 3.1 (ii), we see from Definition 4.1 (iii) that $F_k(T)$ diverges to $-\infty$ as $k \to \infty$. $\qquad \square$

4.4.2 Energy

We prove the following theorem, which leads to the statements about $E(T)$ in Theorem 4.3.

Theorem 4.6 (Tadaki [44]) *Let $T \in \mathbb{R}$.*

(i) *If $0 < T < 1$ and T is computable, then $E(T)$ converges and is a left-computable real which is Chaitin T-random and T-compressible.*

(ii) *If $1 \le T$, then $E(T)$ diverges to ∞.*

Proof (i) Suppose that T is a computable real with $0 < T < 1$.

First we show that $E(T)$ converges. In Definition 4.1 (ii), the denominator $Z_k(T)$ of $E_k(T)$ converges to the real $Z(T) > 0$ as $n \to \infty$. On the other hand, by Theorem 3.4 (i), the numerator $\sum_{i=1}^{k} |q_i| 2^{-|q_i|/T}$ of $E_k(T)$ converges to the real $W(1, T)$ as $n \to \infty$. Thus, $E_k(T)$ converges to the real $W(1, T)/Z(T)$ as $n \to \infty$, and we have

$$E(T) = \frac{W(1, T)}{Z(T)}. \qquad (4.13)$$

By this equality we see that the real $E(T)$ is independent of the choice of the enumeration of $\{q_i\}$ of dom U in Definition 4.1.

Next, we show that $E(T)$ is a left-computable real. Let $p_1, p_2, p_3, \ldots$ be any particular recursive enumeration of the r.e. set $\mathrm{dom}\, U$. For each $m \in \mathbb{N}^+$, we define $\widetilde{W}_m(T)$ and $\widetilde{Z}_m(T)$ by

$$\widetilde{W}_m(T) := \sum_{i=1}^{m} |p_i|\, 2^{-|p_i|/T} \quad \text{and} \quad \widetilde{Z}_m(T) := \sum_{i=1}^{m} 2^{-|p_i|/T},$$

and then define $\widetilde{E}_m(T)$ by $\widetilde{E}_m(T) := \widetilde{W}_m(T)/\widetilde{Z}_m(T)$. Since $\lim_{m \to \infty} \widetilde{W}_m(T) = W(1, T)$ and $\lim_{m \to \infty} \widetilde{Z}_m(T) = Z(T)$, it follows from (4.13) that $\lim_{m \to \infty} \widetilde{E}_m(T) = E(T)$. We then see that

$$\widetilde{E}_{m+1}(T) - \widetilde{E}_m(T) = \frac{\widetilde{Z}_m(T)\, |p_{m+1}| - \widetilde{W}_m(T)}{\widetilde{Z}_{m+1}(T)\widetilde{Z}_m(T)} 2^{-|p_{m+1}|/T}.$$

Since $\widetilde{W}_m(T)$ and $\widetilde{Z}_m(T)$ converge as $m \to \infty$ and $\lim_{m \to \infty} |p_m| = \infty$, there exist $a \in \mathbb{N}^+$ and $m_1 \in \mathbb{N}^+$ such that, for every $m \geq m_1$, it holds that

$$\widetilde{E}_{m+1}(T) - \widetilde{E}_m(T) > |p_{m+1}|\, 2^{-|p_{m+1}|/T - a}. \tag{4.14}$$

In particular, by the above inequality, we see that $\widetilde{E}_m(T)$ is an increasing function of m for all $m \geq m_1$. Thus, since T is a computable real, it follows that $E(T)$ is a left-computable real.

We then show that $E(T)$ is Chaitin T-random. We denote $E(T)$ by α. Since T is a computable real and $\lceil E(T) \rceil - 1 + 0.(\alpha\!\restriction_n^{\mathrm{L}}) < E(T)$ holds for all $n \in \mathbb{N}^+$ due to (2.3), there exists a partial recursive function $\xi \colon \{0, 1\}^* \to \mathbb{N}^+$ such that, for every $n \in \mathbb{N}^+$, it holds that $\xi(\alpha\!\restriction_n^{\mathrm{L}}) \geq m_1$ and

$$\lceil E(T) \rceil - 1 + 0.(\alpha\!\restriction_n^{\mathrm{L}}) < \widetilde{E}_{\xi(\alpha\restriction_n^{\mathrm{L}})}(T).$$

It is then easy to see that

$$E(T) - \widetilde{E}_{\xi(\alpha\restriction_n^{\mathrm{L}})}(T) < 2^{-n}$$

for every $n \in \mathbb{N}^+$. It follows from (4.14) that, for every $n, i \in \mathbb{N}^+$, if $i > \xi(\alpha\!\restriction_n^{\mathrm{L}})$ then $|p_i|\, 2^{-|p_i|/T - a} < 2^{-n}$ and therefore $T \log_2 |p_i| - Ta < |p_i| - Tn$. Thus, given $\alpha\!\restriction_n^{\mathrm{L}}$, by calculating the set $\{ U(p_i) \mid i \leq \xi(\alpha\!\restriction_n^{\mathrm{L}}) \}$ and picking any particular finite binary string which is not in this set, one can obtain an $s \in \{0, 1\}^*$ such that $T \log_2 K(s) - Ta < K(s) - Tn$.

Hence, there exists a partial recursive function $\Psi \colon \{0, 1\}^* \to \{0, 1\}^*$ such that, for every $n \in \mathbb{N}^+$, it holds that

$$T \log_2 K(\Psi(\alpha\!\restriction_n^{\mathrm{L}})) - Ta < K(\Psi(\alpha\!\restriction_n^{\mathrm{L}})) - Tn.$$

Applying this inequality to itself, we have $T \log_2 n < K(\Psi(\alpha\restriction^L_n)) - Tn + O(1)$. On the other hand, using (2.8) there is $c \in \mathbb{N}$ such that, for every $n \in \mathbb{N}^+$, it holds that $K(\Psi(\alpha\restriction^L_n)) < K(\alpha\restriction^L_n) + c$. Therefore, we have

$$T \log_2 n < K(\alpha\restriction^L_n) - Tn + O(1).$$

Hence, $\mathrm{Binary}_L(\alpha)$ is Chaitin T-random. It follows that $\mathrm{Binary}_L(\alpha)$ is not computable, which implies that $\mathrm{Binary}_L(\alpha) = \mathrm{Binary}(\alpha)$. Thus, $E(T)$ is Chaitin T-random.

Next, we show that $E(T)$ is T-compressible. Since T is a computable real, there exists a total recursive function $g \colon \mathbb{N}^+ \times \mathbb{N}^+ \to \mathbb{Z}$ such that, for every $m, n \in \mathbb{N}^+$, it holds that

$$\left| \widetilde{E}_m(T) - \lfloor E(T) \rfloor - 2^{-n} g(m, n) \right| < 2^{-n}. \tag{4.15}$$

It is also shown that there is $h \in \mathbb{N}^+$ such that, for every $m \in \mathbb{N}^+$, it holds that

$$\left| E(T) - \widetilde{E}_m(T) \right| < 2^h \max \left\{ \left| W(1, T) - \widetilde{W}_m(T) \right|, \left| Z(T) - \widetilde{Z}_m(T) \right| \right\}. \tag{4.16}$$

Let t be an arbitrary computable real with $T < t < 1$. Then, $W(1, t) = \lim_{m \to \infty} \widetilde{W}_m(t)$, where

$$\widetilde{W}_m(t) := \sum_{i=1}^{m} |p_i| \, 2^{-|p_i|/t}.$$

It follows from Theorem 3.4 (i) that the limit value $W(1, t)$ exists as a real and is Chaitin t-random. We denote $W(1, t)$ by β.

Given n and $\beta\restriction_{\lceil Tn/t \rceil}$, one can find an $m_0 \in \mathbb{N}^+$ such that

$$\lceil W(1, t) \rceil - 1 + 0.(\beta\restriction_{\lceil Tn/t \rceil}) < \widetilde{W}_{m_0}(t).$$

This is possible since $\lceil W(1, t) \rceil - 1 + 0.(\beta\restriction_{\lceil Tn/t \rceil}) < W(1, t)$ due to (2.3) and $\lim_{m \to \infty} \widetilde{W}_m(t) = W(1, t)$ holds. It is then easy to see that

$$\sum_{i=m_0+1}^{\infty} |p_i| \, 2^{-|p_i|/t} < 2^{-Tn/t}.$$

Raising both sides of this inequality to the power t/T and using the inequality $x^z + y^z \le (x + y)^z$ for reals $x, y > 0$ and $z \ge 1$, it is seen that

$$2^{-n} > \sum_{i=m_0+1}^{\infty} |p_i|^{t/T} \, 2^{-|p_i|/T} > \sum_{i=m_0+1}^{\infty} |p_i| \, 2^{-|p_i|/T}$$

and therefore

$$2^{-n} > \sum_{i=m_0+1}^{\infty} 2^{-|p_i|/T}.$$

It follows that

$$\left| W(1, T) - \widetilde{W}_{m_0}(T) \right| < 2^{-n} \quad \text{and} \quad \left| Z(T) - \widetilde{Z}_{m_0}(T) \right| < 2^{-n}. \tag{4.17}$$

From (4.15), (4.16), (4.17), and $\left| \lfloor E(T) \rfloor + 0.(E(T)\lceil_n) - E(T) \right| < 2^{-n}$, it is shown that

$$\left| E(T)\lceil_n - g(m_0, n) \right| < 2^h + 2$$

and therefore

$$E(T)\lceil_n = g(m_0, n), \; g(m_0, n) \pm 1, \; g(m_0, n) \pm 2, \; \ldots, \; g(m_0, n) \pm (2^h + 1),$$

where $E(T)\lceil_n$ is regarded as a dyadic integer. Thus, one is left with $2^{h+1} + 3$ possibilities of $E(T)\lceil_n$, so that one needs only $h + 2$ bits more in order to determine $E(T)\lceil_n$.

Thus, there exists a partial recursive function $\Phi \colon \mathbb{N}^+ \times \{0, 1\}^* \times \{0, 1\}^* \to \{0, 1\}^*$ such that for every $n \in \mathbb{N}^+$ there exists $s \in \{0, 1\}^{h+2}$ with the property that $\Phi(n, \beta\lceil_{\lceil Tn/t \rceil}, s) = E(T)\lceil_n$. Consider a prefix-free machine M such that, for every $p, v \in \{0, 1\}^*$, $M(p) = v$ if and only if there exist $q \in \mathrm{dom}\, U, r \in \{0, 1\}^*, s \in \{0, 1\}^{h+2}$, and $n \in \mathbb{N}^+$ with the properties that $p = qrs$, $U(q) = n$, $|r| = \lceil Tn/t \rceil$, and $\Phi(n, r, s) = v$. Note that such a prefix-free machine M exists since T and t are computable. Then, it is easy to see that

$$K_M(E(T)\lceil_n) \leq K(n) + |\beta\lceil_{\lceil Tn/t \rceil}| + h + 2$$

for every $n \in \mathbb{N}^+$. Thus, since $\lim_{n \to \infty} K(n)/n = 0$ due to (2.7), and also $|\beta\lceil_{\lceil Tn/t \rceil}| \leq Tn/t + 1$ holds, it follows from (2.4) that $E(T)$ is T/t-compressible. Since t is an arbitrary computable real with $T < t < 1$, it follows that $E(T)$ is T-compressible.

(ii) In the case of $T = 1$, by Theorem 3.4 (ii), the numerator

$$W(1, 1) = \sum_{p \in \mathrm{dom}\, U} |p| 2^{-|p|}$$

of $E(1)$ diverges to ∞. On the other hand, the denominator $Z(1)$ of $E(1)$ converges. Thus, $E(1)$ diverges to ∞.

The case of $T > 1$ is treated as follows. We note that $\lim_{k \to \infty} |q_k| = \infty$. Given $M > 0$, there is $k_0 \in \mathbb{N}^+$ such that $|q_i| \geq 2M$ for all $i > k_0$. Since $\lim_{k \to \infty} Z_k(T) = \infty$ by Theorem 3.1 (ii), there is $k_1 \in \mathbb{N}^+$ such that

$$\frac{1}{Z_k(T)} \sum_{i=1}^{k_0} 2^{-|q_i|/T} \leq \frac{1}{2}$$

for all $k > k_1$. Thus, for each $k > \max\{k_0, k_1\}$,

$$E_k(T) = \frac{1}{Z_k(T)} \sum_{i=1}^{k_0} |q_i| 2^{-|q_i|/T} + \frac{1}{Z_k(T)} \sum_{i=k_0+1}^{k} |q_i| 2^{-|q_i|/T}$$

$$> \frac{2M}{Z_k(T)} \sum_{i=k_0+1}^{k} 2^{-|q_i|/T}$$

$$= 2M \left(1 - \frac{1}{Z_k(T)} \sum_{i=1}^{k_0} 2^{-|q_i|/T} \right)$$

$$\geq 2M \frac{1}{2} = M.$$

Hence, $\lim_{k \to \infty} E_k(T) = \infty$. This completes the proof. $\qquad\qquad\square$

4.4.3 Entropy

We prove the following theorem, which leads to the statements about $S(T)$ in Theorem 4.3.

Theorem 4.7 (Tadaki [44]) *Let $T \in \mathbb{R}$.*

(i) *If $0 < T < 1$ and T is computable, then $S(T)$ converges and is a left-computable real which is Chaitin T-random and T-compressible.*

(ii) *If $1 \leq T$, then $S(T)$ diverges to ∞.*

Proof (i) Suppose that T is a computable real with $0 < T < 1$.

Since $Z(T)$ and $E(T)$ converge by Theorem 3.1 (i) and Theorem 4.6 (i), respectively, it follows from Definition 4.1 that $S(T)$ also converges and satisfies that

$$S(T) = \frac{1}{T} E(T) + \log_2 Z(T). \qquad (4.18)$$

By this equality we see that the real $S(T)$ is independent of the choice of the enumeration of $\{q_i\}$ of dom U in Definition 4.1.

Since $E(T)$ is a left-computable real and $E(T) \notin \mathbb{N}$ by Theorem 4.6 (i), there exists a total recursive function $f : \mathbb{N}^+ \to \mathbb{Q}$ such that $0 < f(m) \leq E(T)$ and $\lfloor f(m) \rfloor = \lfloor E(T) \rfloor$ for all $m \in \mathbb{N}^+$ and $\lim_{m \to \infty} f(m) = E(T)$. Since T is a computable real, there exists a total recursive function $g : \mathbb{N}^+ \to \mathbb{Q}$ such that $0 \leq g(m) \leq 1/T$ for all $m \in \mathbb{N}^+$ and $\lim_{m \to \infty} g(m) = 1/T$. Moreover, since $Z(T)$ is a left-computable real by Theorem 4.4, there exists a total recursive function $h : \mathbb{N}^+ \to \mathbb{Q}$ such that $h(m) \leq \log_2 Z(T)$ for all $m \in \mathbb{N}^+$ and $\lim_{m \to \infty} h(m) = \log_2 Z(T)$. Hence, $g(m) f(m) + h(m) \leq S(T)$ for all $m \in \mathbb{N}^+$ and

$$\lim_{m \to \infty} [g(m) f(m) + h(m)] = S(T). \qquad (4.19)$$

Thus, $S(T)$ is a left-computable real.

We then show that $S(T)$ is Chaitin T-random. We denote $S(T)$ by α.
Given $\alpha|_n^L$, one can find an $m_0 \in \mathbb{N}^+$ such that

$$\lceil S(T) \rceil - 1 + 0.(\alpha|_n^L) < g(m_0)f(m_0) + h(m_0).$$

This is possible because $\lceil S(T) \rceil - 1 + 0.(\alpha|_n^L) < S(T)$ due to (2.3), and (4.19) holds. From (2.3), it is shown that

$$2^{-n} > \left(\frac{1}{T}E(T) + \log_2 Z(T) \right) - (g(m_0)f(m_0) + h(m_0))$$
$$\geq \frac{1}{T}E(T) - g(m_0)f(m_0) \geq \frac{1}{T}(E(T) - f(m_0))$$
$$\geq E(T) - f(m_0).$$

Thus, $0 \leq E(T) - f(m_0) < 2^{-n}$. One then calculates $f(m_0)|_n$. It follows from (2.2) that

$$0 \leq 0.(E(T)|_n) - 0.l_n < E(T) - f(m_0) + 2^{-n} < 2 \cdot 2^{-n},$$

where $f(m_0)|_n$ is denoted by l_n. Hence $E(T)|_n = l_n,\ l_n + 1$, where $E(T)|_n$ and l_n are regarded as a dyadic integer. Thus, one is left with only two possibilities of $E(T)|_n$, so that one needs only 1 bit more in order to determine $E(T)|_n$.

Thus, there exists a partial recursive function $\Phi : \{0,1\}^* \times \{0,1\}^* \to \{0,1\}^*$ such that for every $n \in \mathbb{N}^+$ there exists $b \in \{0,1\}$ for which $\Phi(\alpha|_n^L, b) = E(T)|_n$ holds. Consider a prefix-free machine M such that, for every $p, v \in \{0,1\}^*$, $M(p) = v$ if and only if there exist $q \in \text{dom}\,U$, $b \in \{0,1\}$, and $s \in \{0,1\}^*$ with the properties that $p = qb$, $U(q) = s$, and $\Phi(s, b) = v$. Note that such a prefix-free machine M exists. Then, it is easy to see that $K_M(E(T)|_n) \leq K(\alpha|_n^L) + 1$ for every $n \in \mathbb{N}^+$. It follows from (2.4) that $K(E(T)|_n) \leq K(\alpha|_n^L) + O(1)$. Thus, since $E(T)$ is Chaitin T-random by Theorem 4.6 (i), $\text{Binary}_L(\alpha)$ is also Chaitin T-random. It follows that $\text{Binary}_L(\alpha)$ is not computable, which implies that $\text{Binary}_L(\alpha) = \text{Binary}(\alpha)$. Hence, $S(T)$ is Chaitin T-random.

Next, we show that $S(T)$ is T-compressible. Let $p_1, p_2, p_3, \ldots$ be any particular recursive enumeration of the r.e. set $\text{dom}\,U$. For each $m \in \mathbb{N}^+$, we define $\widetilde{W}_m(T)$ and $\widetilde{Z}_m(T)$ by

$$\widetilde{W}_m(T) := \sum_{i=1}^{m} |p_i| 2^{-|p_i|/T} \quad \text{and} \quad \widetilde{Z}_m(T) := \sum_{i=1}^{m} 2^{-|p_i|/T},$$

and then define $\widetilde{S}_m(T)$ by

$$\widetilde{S}_m(T) := \frac{1}{T}\frac{\widetilde{W}_m(T)}{\widetilde{Z}_m(T)} + \log_2 \widetilde{Z}_m(T).$$

Since $\lim_{m\to\infty}\widetilde{W}_m(T) = W(1, T)$ and $\lim_{m\to\infty}\widetilde{Z}_m(T) = Z(T)$, it follows from (4.18) and $E(T) = W(1, T)/Z(T)$ that $\lim_{m\to\infty}\widetilde{S}_m(T) = S(T)$.

Since T is a computable real, there exists a total recursive function $a\colon \mathbb{N}^+ \times \mathbb{N}^+ \to \mathbb{Z}$ such that, for every $m, n \in \mathbb{N}^+$, it holds that

$$\left|\, \widetilde{S}_m(T) - \lfloor S(T)\rfloor - 2^{-n}a(m, n)\,\right| < 2^{-n}. \tag{4.20}$$

It is also shown that there is $d \in \mathbb{N}^+$ such that, for every $m \in \mathbb{N}^+$, it holds that

$$\left|\, S(T) - \widetilde{S}_m(T)\,\right| < 2^d \max\left\{\left|\,W(1, T) - \widetilde{W}_m(T)\,\right|, \left|\,Z(T) - \widetilde{Z}_m(T)\,\right|\right\}. \tag{4.21}$$

Based on the inequalities (4.20) and (4.21), in the same manner as the proof of the T-compressibility of $E(T)$ in Theorem 4.6 (i), we can show that $S(T)$ is T-compressible.

(ii) Note from Definition 4.1 that

$$S_k(T) = \frac{1}{T}E_k(T) + \log_2 Z_k(T)$$

for every $T > 0$ and $k \in \mathbb{N}^+$. Thus, in the case of $T \geq 1$, since $\lim_{k\to\infty} E_k(T)=\infty$ by Theorem 4.6 (ii) and $\log_2 Z_k(T)$ is bounded to the below, we see that $\lim_{k\to\infty} S_k(T) = \infty$. $\qquad\qquad\square$

4.4.4 Specific Heat

We prove the following theorem, which leads to the statements about $C(T)$ in Theorem 4.3.

Theorem 4.8 (Tadaki [44]) *Let $T \in \mathbb{R}$.*

(i) *If $0 < T < 1$ and T is computable, then $C(T)$ converges and is a left-computable real which is Chaitin T-random and T-compressible, and moreover $C(T) = E'(T)$ where $E'(T)$ is the derived function of $E(T)$.*

(ii) *If $T = 1$, then $C(T)$ diverges to ∞.*

Proof (i) Suppose that T is a computable real with $0 < T < 1$.

First we show that $C(T)$ converges. It is easy to show that, for every $k \in \mathbb{N}^+$, it holds that

$$C_k(T) = \frac{\ln 2}{T^2}\left\{\frac{Y_k(T)}{Z_k(T)} - \left(\frac{W_k(T)}{Z_k(T)}\right)^2\right\},$$

where

$$Y_k(T) := \sum_{i=1}^{k}|q_i|^2\, 2^{-|q_i|/T} \quad\text{and}\quad W_k(T) := \sum_{i=1}^{k}|q_i|\, 2^{-|q_i|/T}.$$

Note that $Z_k(T)$ converges to the real $Z(T) > 0$ as $k \to \infty$. On the other hand, by Theorem 3.4 (i), $Y_k(T)$ and $W_k(T)$ converge to the reals $W(2, T)$ and $W(1, T)$, respectively, as $k \to \infty$. Thus, $C_k(T)$ also converges to a real $C(T)$ as $k \to \infty$, and we have

$$C(T) = \frac{\ln 2}{T^2} \left\{ \frac{W(2, T)}{Z(T)} - \left(\frac{W(1, T)}{Z(T)} \right)^2 \right\}. \tag{4.22}$$

By this equality we see that the real $C(T)$ is independent of the choice of the enumeration of $\{q_i\}$ of dom U in Definition 4.1.

Next, we show that $C(T)$ is a left-computable real. Let $p_1, p_2, p_3, \ldots$ be any particular recursive enumeration of the r.e. set dom U. For each $m \in \mathbb{N}^+$, we define $\widetilde{Y}_m(T)$, $\widetilde{W}_m(T)$, and $\widetilde{Z}_m(T)$ by

$$\widetilde{Y}_m(T) := \sum_{i=1}^{m} |p_i|^2 \, 2^{-|p_i|/T}, \quad \widetilde{W}_m(T) := \sum_{i=1}^{m} |p_i| \, 2^{-|p_i|/T}, \quad \text{and} \quad \widetilde{Z}_m(T) := \sum_{i=1}^{m} 2^{-|p_i|/T},$$

respectively, and then define $\widetilde{C}_m(T)$ by

$$\widetilde{C}_m(T) := \frac{\ln 2}{T^2} \left\{ \frac{\widetilde{Y}_m(T)}{\widetilde{Z}_m(T)} - \left(\frac{\widetilde{W}_m(T)}{\widetilde{Z}_m(T)} \right)^2 \right\}.$$

Since $\lim_{m \to \infty} \widetilde{Y}_m(T) = W(2, T)$, $\lim_{m \to \infty} \widetilde{W}_m(T) = W(1, T)$, and $\lim_{m \to \infty} \widetilde{Z}_m(T) = Z(T)$, it follows from (4.22) that $\lim_{m \to \infty} \widetilde{C}_m(T) = C(T)$. We then see that $\widetilde{C}_{m+1}(T) - \widetilde{C}_m(T)$ is calculated as

$$\frac{\ln 2}{T^2} \frac{2^{-|p_{m+1}|/T}}{\widetilde{Z}_{m+1}(T)} \left[|p_{m+1}|^2 - \left\{ \frac{\widetilde{W}_{m+1}(T)}{\widetilde{Z}_{m+1}(T)} + \frac{\widetilde{W}_m(T)}{\widetilde{Z}_m(T)} \right\} |p_{m+1}| \right. $$
$$\left. + \left\{ \frac{\widetilde{W}_{m+1}(T)}{\widetilde{Z}_{m+1}(T)} + \frac{\widetilde{W}_m(T)}{\widetilde{Z}_m(T)} \right\} \frac{\widetilde{W}_m(T)}{\widetilde{Z}_m(T)} - \frac{\widetilde{Y}_m(T)}{\widetilde{Z}_m(T)} \right]. \tag{4.23}$$

Since $\widetilde{Y}_m(T)$, $\widetilde{W}_m(T)$, and $\widetilde{Z}_m(T)$ converge as $m \to \infty$ and $\lim_{m \to \infty} |p_m| = \infty$, there exists $m_1 \in \mathbb{N}^+$ such that, for every $m \geq m_1$, it holds that

$$\widetilde{C}_{m+1}(T) - \widetilde{C}_m(T) > |p_{m+1}| \, 2^{-|p_{m+1}|/T}. \tag{4.24}$$

In particular, by the above inequality, we see that $\widetilde{C}_m(T)$ is an increasing function of m for all $m \geq m_1$. Thus, since T is a computable real, it follows that $C(T)$ is a left-computable real.

Based on the computability of T and the inequality (4.24), in the same manner as the proof of Theorem 4.6 (i), we can show that $C(T)$ is Chaitin T-random.

Next, we show that $C(T)$ is T-compressible. Since T is a computable real, there exists a total recursive function $g: \mathbb{N}^+ \times \mathbb{N}^+ \to \mathbb{Z}$ such that, for every $m, n \in \mathbb{N}^+$, it holds that

$$\left| \tilde{C}_m(T) - \lfloor C(T) \rfloor - 2^{-n}g(m,n) \right| < 2^{-n}. \tag{4.25}$$

It is also shown that there is $c \in \mathbb{N}^+$ such that, for every $m \in \mathbb{N}^+$, it holds that

$$\left| C(T) - \tilde{C}_m(T) \right|$$
$$< 2^c \max \left\{ \left| Y(T) - \tilde{Y}_m(T) \right|, \left| W(T) - \tilde{W}_m(T) \right|, \left| Z(T) - \tilde{Z}_m(T) \right| \right\}. \tag{4.26}$$

Let t be an arbitrary computable real with $T < t < 1$. Then, $W(2,t) = \lim_{m \to \infty} \tilde{Y}_m(t)$, where

$$\tilde{Y}_m(t) := \sum_{i=1}^{m} |p_i|^2 \, 2^{-|p_i|/t}.$$

It follows from Theorem 3.4 (i) that the limit value $W(2,t)$ exists as a real and is Chaitin t-random. We denote $W(2,t)$ by β.

Given n and $\beta\restriction_{\lceil Tn/t \rceil}$, one can find an $m_0 \in \mathbb{N}^+$ such that

$$\lceil W(2,t) \rceil - 1 + 0.(\beta\restriction_{\lceil Tn/t \rceil}) < \tilde{Y}_{m_0}(t).$$

This is possible since $\lceil W(2,t) \rceil - 1 + 0.(\beta\restriction_{\lceil Tn/t \rceil}) < W(2,t)$ due to (2.3) and $\lim_{m \to \infty} \tilde{Y}_m(t) = W(2,t)$ holds. It is then easy to see that

$$\sum_{i=m_0+1}^{\infty} |p_i|^2 \, 2^{-|p_i|/t} < 2^{-Tn/t}.$$

Raising both sides of this inequality to the power t/T and using the inequality $x^z + y^z \le (x+y)^z$ for reals $x, y > 0$ and $z \ge 1$, it is seen that

$$2^{-n} > \sum_{i=m_0+1}^{\infty} |p_i|^{2t/T} \, 2^{-|p_i|/T} > \sum_{i=m_0+1}^{\infty} |p_i|^2 \, 2^{-|p_i|/T}$$

and therefore

$$2^{-n} > \sum_{i=m_0+1}^{\infty} |p_i| \, 2^{-|p_i|/T} \quad \text{and} \quad 2^{-n} > \sum_{i=m_0+1}^{\infty} 2^{-|p_i|/T}.$$

It follows that

$$\max \left\{ \left| W(2,T) - \tilde{Y}_{m_0}(T) \right|, \left| W(1,T) - \tilde{W}_{m_0}(T) \right|, \left| Z(T) - \tilde{Z}_{m_0}(T) \right| \right\} < 2^{-n}. \tag{4.27}$$

From (4.25), (4.26), (4.27), and $\left| \lfloor C(T) \rfloor + 0.(C(T)\restriction_n) - C(T) \right| < 2^{-n}$, it is shown that $\left| C(T)\restriction_n - g(m_0, n) \right| < 2^c + 2$ and therefore

$$C(T)\restriction_n = g(m_0, n), \ g(m_0, n) \pm 1, \ g(m_0, n) \pm 2, \ \ldots, \ g(m_0, n) \pm (2^c + 1),$$

where $C(T)\restriction_n$ is regarded as a dyadic integer. Thus, one is left with $2^{c+1} + 3$ possibilities of $C(T)\restriction_n$, so that one needs only $c + 2$ bits more in order to determine $C(T)\restriction_n$.

Thus, there exists a partial recursive function $\Phi: \mathbb{N}^+ \times \{0, 1\}^* \times \{0, 1\}^* \to \{0, 1\}^*$ such that for every $n \in \mathbb{N}^+$ there exists $s \in \{0, 1\}^{c+2}$ with the property that $\Phi(n, \beta\restriction_{\lceil Tn/t \rceil}, s) = C(T)\restriction_n$. Consider a prefix-free machine M such that, for every $p, v \in \{0, 1\}^*$, $M(p) = v$ if and only if there exist $q \in \operatorname{dom} U$, $r \in \{0, 1\}^*$, $s \in \{0, 1\}^{c+2}$, and $n \in \mathbb{N}^+$ with the properties that $p = qrs$, $U(q) = n$, $|r| = \lceil Tn/t \rceil$, and $\Phi(n, r, s) = v$. Note that such a prefix-free machine M exists since T and t are computable. Then, it is easy to see that

$$K_M(C(T)\restriction_n) \le K(n) + |\beta\restriction_{\lceil Tn/t \rceil}| + c + 2$$

for every $n \in \mathbb{N}^+$. Thus, since $\lim_{n \to \infty} K(n)/n = 0$ due to (2.7), and also $|\beta\restriction_{\lceil Tn/t \rceil}| \le Tn/t + 1$ holds, it follows from (2.4) that $C(T)$ is T/t-compressible. Since t is an arbitrary computable real with $T < t < 1$, it follows that $C(T)$ is T-compressible.

Finally, by evaluating $C_{k+1}(x) - C_k(x)$ for all $x \in (0, 1)$ like (4.23), we can show that $C_k(x)$ converges uniformly in the wider sense on $(0, 1)$ to $C(x)$ as $k \to \infty$. Hence, we have $C(T) = E'(T)$.

(ii) It can be shown that

$$C_k(1) = \frac{\ln 2}{2} \frac{1}{Z_k(1)^2} \sum_{i=1}^{k} \sum_{j=1}^{k} (|q_i| - |q_j|)^2 2^{-|q_i|} 2^{-|q_j|}.$$

By Theorem 3.5 (i), $\sum_{j=1}^{k} (|q_1| - |q_j|)^2 2^{-|q_j|}$ diverges to ∞ as $k \to \infty$. Therefore $\sum_{i=1}^{k} \sum_{j=1}^{k} (|q_i| - |q_j|)^2 2^{-|q_i|} 2^{-|q_j|}$ also diverges to ∞ as $k \to \infty$. On the other hand, $Z_k(1)$ converges to Ω as $k \to \infty$. Thus, $C(1)$ diverges to ∞. This completes the proof. $\qquad \square$

4.5 A Privileged Status of Thermodynamic Quantities of AIT

Theorems 4.2 and 4.3 show that the partial randomness (and therefore the compression rate) of the values of all the thermodynamic quantities $Z(T), E(T), F(T), S(T)$, and $C(T)$ in Definition 4.1 equals to the temperature T if T is a computable real with $0 < T < 1$. The Boltzmann factor $2^{-|p|/T}$ appears in the definitions of all of the thermodynamic quantities of AIT. We can see, for example, that

$$E(T) = \frac{\sum_{i=1}^{\infty} |q_i| \, 2^{-|q_i|/T}}{\sum_{i=1}^{\infty} 2^{-|q_i|/T}},$$

$$C(T) = \frac{d}{dT} E(T) = \frac{\ln 2}{T^2} \left\{ \frac{\sum_{i=1}^{\infty} |q_i|^2 \, 2^{-|q_i|/T}}{\sum_{i=1}^{\infty} 2^{-|q_i|/T}} - \left(\frac{\sum_{i=1}^{\infty} |q_i| \, 2^{-|q_i|/T}}{\sum_{i=1}^{\infty} 2^{-|q_i|/T}} \right)^2 \right\}$$

for each real T with $0 < T < 1$. Actually, the proofs of Theorems 4.2 and 4.3 show the importance of the role of the Boltzmann factor $2^{-|p|/T}$ in deriving them.

Note, however, that the partial randomness of every function of T involving the Boltzmann factor $2^{-|p|/T}$ does not necessarily equal to T. To see this, consider the following quantity $\overline{Z}(T)$, which is "unnatural" from the point of view of statistical mechanics.

$$\overline{Z}(T) := \sum_{i=1}^{\infty} \left(2^{-|q_i|/T} \right)^2.$$

Since $\overline{Z}(T) = Z(T/2)$, it follows from Theorem 4.2 (i) that if T is a computable real with $0 < T < 2$, then $\overline{Z}(T)$ converges and its partial randomness does not equal to T but equals to $T/2$. The results of this chapter, together with this counter-example, reveal one of the aspects of a privileged status of the thermodynamic quantities of AIT.

In this chapter we have shown that the partial randomness of the values of all the thermodynamic quantities of AIT equals to the temperature T. In the next chapter we will show that this situation holds for the *temperature itself* as a thermodynamic quantity. Specifically, we will show that the computability of the value of the thermodynamic quantities of AIT at temperature $T \in (0, 1)$ gives a sufficient condition for T to be a fixed point on partial randomness. This result of the next chapter reveals another aspect of a privileged status of the thermodynamic quantities of AIT.

Chapter 5
Fixed Point Theorems on Partial Randomness

5.1 Self-Referential Nature of Temperature

In the preceding chapter, we saw that the partial randomness (and therefore the compression rate) of the values of all the thermodynamic quantities equals to the temperature T in the statistical mechanical interpretation of AIT. However, in statistical mechanics or thermodynamics, among all thermodynamic quantities one of the most typical thermodynamic quantities is *temperature* itself. Inspired by this fact, the following question thus arises naturally.

Question: Can the partial randomness of the temperature T equal the temperature T itself in the statistical mechanical interpretation of AIT?

This question is rather self-referential. However, we can answer it *affirmatively* in the following form.

Theorem 5.1 (Fixed Point Theorem on Partial Randomness, Tadaki [44]) *For every real T with $0 < T < 1$, if the value of the partition function $Z(T)$ at temperature T is a computable real, then the real T is weakly Chaitin T-random and T-compressible, and therefore the compression rate of the real T equals to T, i.e., we have*

$$\lim_{n \to \infty} \frac{K(T\restriction_n)}{n} = T. \tag{5.1}$$

$\square$

Theorem 5.1 follows immediately from Theorem 5.3, Theorem 5.4, and Theorem 5.5 below.

Intuitively, we might interpret the meaning of (5.1) as follows: Imagine a file of infinite size whose content is

"The compression rate of this file is $0.100111001 \ldots \ldots$"

This chapter is based on Tadaki [44, 52], and is a rearrangement of the parts of them.

© The Author(s), under exclusive license to Springer Nature Singapore Pte Ltd. 2019
K. Tadaki, *A Statistical Mechanical Interpretation of Algorithmic Information Theory*,
SpringerBriefs in Mathematical Physics 36,
https://doi.org/10.1007/978-981-15-0739-7_5

When this file is compressed, the compression rate of this file is actually equal to

$$0.100111001\ldots\ldots,$$

as the content of this file says. This situation is self-referential. Since it holds in Theorem 5.1 that the partial randomness (and therefore the compression rate) of the real T equals to T, the real T is a *fixed point on partial randomness* (and therefore a *fixed point on compression rate*). Thus, from a purely mathematical point of view, Theorem 5.1 is just a *fixed point theorem on partial randomness*.

In Theorem 5.1, the computability of the value $Z(T)$ gives a *sufficient* condition for a real T to be a fixed point on partial randomness. The following theorem holds for this sufficient condition.

Theorem 5.2 (Tadaki [44]) *The set* $\{T \in (0, 1) \mid Z(T)$ *is computable*$\}$ *is dense in* $(0, 1)$.

Proof Since the function $2^{-l/T}$ of T is increasing on $(0, 1)$ for each $l \in \mathbb{N}^+$, the function $Z(T)$ of T is increasing on $(0, 1)$. On the other hand, the function $Z(T)$ of T is continuous on $(0, 1)$.[1] Actually, Tadaki [40, 41] showed that $Z(T)$ is a function of class C^∞ on $(0, 1)$. Thus, since the set of all computable reals is dense in $\mathbb{R}$, the result follows.   □

Thus, we have the following corollary of Theorem 5.1.

Corollary 5.1 *The set* $\{T \in (0, 1) \mid$ *The compression rate of* T *equals to* $T\}$ *is dense in* $(0, 1)$.   □

From the point of view of the statistical mechanical interpretation of AIT introduced in the preceding chapter, Theorem 5.1 shows that the partial randomness of temperature is equal to the temperature itself. Thus, Theorem 5.1 further confirms the role of temperature as the partial randomness of the thermodynamic quantities of AIT, which we observed in the preceding chapter.

Now, let us prove Theorem 5.1. As a first step to prove it, we prove the following theorem which gives the weak Chaitin T-randomness of T in Theorem 5.1.

Theorem 5.3 *For every* $T \in (0, 1)$, *if* $Z(T)$ *is a right-computable real, then* T *is weakly Chaitin* T-*random.*

Proof Let $p_1, p_2, p_3, \ldots$ be any particular recursive enumeration of the r.e. set dom U. For each $k \in \mathbb{N}^+$, we define a function $\widetilde{Z}_k \colon (0, 1) \to \mathbb{R}$ by

$$\widetilde{Z}_k(x) = \sum_{i=1}^{k} 2^{-|p_i|/x}.$$

Then, $\lim_{k \to \infty} \widetilde{Z}_k(x) = Z(x)$ for every $x \in (0, 1)$.

---

[1] We will give a proof of this fact in Lemma 5.1 (iii) in Sect. 5.2 later.

Suppose that $T \in (0, 1)$ and $Z(T)$ is right-computable. Then there exists a total recursive function $g \colon \mathbb{N}^+ \to \mathbb{Q}$ such that $Z(T) \le g(m)$ for all $m \in \mathbb{N}^+$, and $\lim_{m \to \infty} g(m) = Z(T)$. We choose any particular real t with $T < t < 1$. For each $i \in \mathbb{N}^+$, using the mean value theorem we see that

$$2^{-|p_i|/x} - 2^{-|p_i|/T} < \frac{\ln 2}{T^2} |p_i| \, 2^{-|p_i|/t}(x - T)$$

for all $x \in (T, t)$. We choose any particular $c \in \mathbb{N}$ with $W(1, t) \ln 2/T^2 \le 2^c$. Here, the limit value $W(1, t)$ exists as a real by Theorem 3.4 (i), since $0 < t < 1$. Then, it follows that

$$\tilde{Z}_k(x) - \tilde{Z}_k(T) < 2^c(x - T) \tag{5.2}$$

for all $k \in \mathbb{N}^+$ and $x \in (T, t)$.

We choose any particular $n_0 \in \mathbb{N}^+$ such that $T < 0.(T \restriction_n) + 2^{-n} < t$ for all $n \ge n_0$. Such n_0 exists since $T < t$ and $\lim_{n \to \infty} 0.(T \restriction_n) + 2^{-n} = T$.

Given $T \restriction_n$ with $n \ge n_0$, one can find $k_0, m_0 \in \mathbb{N}^+$ such that

$$g(m_0) < \tilde{Z}_{k_0}(0.(T \restriction_n) + 2^{-n}).$$

This is possible from $Z(T) < Z(0.(T \restriction_n) + 2^{-n})$,

$$\lim_{k \to \infty} \tilde{Z}_k(0.(T \restriction_n) + 2^{-n}) = Z(0.(T \restriction_n) + 2^{-n}),$$

and the properties of the function g. It follows from $Z(T) \le g(m_0)$ and (5.2) that

$$\sum_{i=k_0+1}^{\infty} 2^{-|p_i|/T} = Z(T) - \tilde{Z}_{k_0}(T) < \tilde{Z}_{k_0}(0.(T \restriction_n) + 2^{-n}) - \tilde{Z}_{k_0}(T) < 2^{c-n}.$$

Hence, for every $i > k_0$, $2^{-|p_i|/T} < 2^{c-n}$ and therefore $Tn - Tc < |p_i|$. Thus, by calculating the set $\{ U(p_i) \mid i \le k_0 \}$ and picking any particular finite binary string which is not in this set, one can then obtain an $s \in \{0, 1\}^*$ such that $Tn - Tc < K(s)$.

Hence, there exists a partial recursive function $\Psi \colon \{0, 1\}^* \to \{0, 1\}^*$ such that $Tn - Tc < K(\Psi(T \restriction_n))$ for all $n \ge n_0$. On the other hand, using (2.8) there is $d \in \mathbb{N}$ such that $K(\Psi(T \restriction_n)) < K(T \restriction_n) + d$ for all $n \ge n_0$. Therefore,

$$Tn - Tc - d < K(T \restriction_n)$$

for all $n \ge n_0$. It follows that T is weakly Chaitin T-random.   □

Remark 5.1 By elaborating Theorem 3.1 (i), we can see that the left-computability of T results in the weak Chaitin T-randomness of $Z(T)$. On the other hand, by Theorem 5.3, the right-computability of $Z(T)$ results in the weak Chaitin T-randomness of T. We can integrate these two extremes into the following form:

For every $T \in (0, 1]$, there exists $c \in \mathbb{N}^+$ such that, for every $n \in \mathbb{N}^+$, it holds that

$$Tn - c \leq K(T{\upharpoonright}_n, Z(T){\upharpoonright}_n). \tag{5.3}$$

In fact, if T is left-computable, then we can show that

$$K(Z(T){\upharpoonright}_n) = K(T{\upharpoonright}_n, Z(T){\upharpoonright}_n) + O(1),$$

and therefore the inequality (5.3) results in the weak Chaitin T-randomness of $Z(T)$. On the other hand, if $Z(T)$ is right-computable, then we can show that

$$K(T{\upharpoonright}_n) = K(T{\upharpoonright}_n, Z(T){\upharpoonright}_n) + O(1),$$

and therefore the inequality (5.3) results in the weak Chaitin T-randomness of T.

Note, however, that the inequality (5.3) is not necessarily tight except for these two extremes. In other words, the following inequality does not hold: For every $T \in (0, 1]$,

$$K(T{\upharpoonright}_n, Z(T){\upharpoonright}_n) \leq Tn + o(n), \tag{5.4}$$

where $o(n)$ depends on T in addition to n. To see this, contrarily assume that the inequality (5.4) holds. Then, by setting T to Chaitin's Ω, we have

$$K(\Omega{\upharpoonright}_n) \leq K(\Omega{\upharpoonright}_n, Z(\Omega){\upharpoonright}_n) + O(1) \leq \Omega n + o(n).$$

Since $\Omega < 1$, this contradicts the fact that Ω is weakly Chaitin random. Thus, the inequality (5.4) does not hold. $\qquad\square$

Theorem 5.4 and Theorem 5.5 below give the T-compressibility of T in Theorem 5.1 together.

Theorem 5.4 *For every $T \in (0, 1)$, if $Z(T)$ is a right-computable real, then T is also a right-computable real.*

Proof Let $p_1, p_2, p_3, \ldots$ be any particular recursive enumeration of the r.e. set $\operatorname{dom} U$. For each $k \in \mathbb{N}^+$, we define a function $\widetilde{Z}_k \colon (0, 1) \to \mathbb{R}$ by

$$\widetilde{Z}_k(x) = \sum_{i=1}^{k} 2^{-|p_i|/x}.$$

Then, $\lim_{k\to\infty} \widetilde{Z}_k(x) = Z(x)$ for every $x \in (0, 1)$.

Suppose that $T \in (0, 1)$ and $Z(T)$ is right-computable. Then there exists a total recursive function $g \colon \mathbb{N}^+ \to \mathbb{Q}$ such that $Z(T) \leq g(m)$ for all $m \in \mathbb{N}^+$, and $\lim_{m\to\infty} g(m) = Z(T)$. Since $Z(x)$ is an increasing function of $x \in (0, 1)$, we see that, for every $x \in \mathbb{Q}$ with $0 < x < 1$, $T < x$ if and only if there are $m, k \in \mathbb{N}^+$ such that $g(m) < \widetilde{Z}_k(x)$. Thus, T is right-computable. This is because the set

$$\{(m,n) \mid m \in \mathbb{Z} \ \& \ n \in \mathbb{N}^+ \ \& \ T < m/n\}$$

is r.e. if and only if T is right-computable.   □

The converse of Theorem 5.4 does not hold. To see this, consider an arbitrary computable real $T \in (0,1)$. Then, obviously T is right-computable. On the other hand, $Z(T)$ is left-computable and weakly Chaitin T-random by Theorem 4.4 and Theorem 3.1 (i), respectively. Thus, $Z(T)$ is not right-computable.

Theorem 5.5 *For every $T \in (0,1)$, if $Z(T)$ is a left-computable real and T is a right-computable real, then T is T-compressible.*

Proof Let $p_1, p_2, p_3, \ldots$ be any particular recursive enumeration of the r.e. set dom U. For each $k \in \mathbb{N}^+$, we define a function $\tilde{Z}_k : (0,1) \to \mathbb{R}$ by

$$\tilde{Z}_k(x) = \sum_{i=1}^{k} 2^{-|p_i|/x}.$$

Then, $\lim_{k\to\infty} \tilde{Z}_k(x) = Z(x)$ for every $x \in (0,1)$.

Suppose that T is a right-computable real with $T \in (0,1)$ and $Z(T)$ is left-computable. For each $i \in \mathbb{N}^+$, using the mean value theorem we see that

$$2^{-|p_1|/t} - 2^{-|p_1|/T} > (\ln 2)\,|p_1|\,2^{-|p_1|/T}\,(t - T)$$

for all $t \in (T,1)$. We choose any particular $c \in \mathbb{N}^+$ such that $(\ln 2)\,|p_1|\,2^{-|p_1|/T} \geq 2^{-c}$. Then, it follows that

$$\tilde{Z}_k(t) - \tilde{Z}_k(T) > 2^{-c}(t - T) \tag{5.5}$$

for all $k \in \mathbb{N}^+$ and $t \in (T,1)$.

Since T is a right-computable real with $T < 1$, there exists a total recursive function $f : \mathbb{N}^+ \to \mathbb{Q}$ such that $T < f(l) < 1$ for all $l \in \mathbb{N}^+$, and $\lim_{l\to\infty} f(l) = T$. On the other hand, since $Z(T)$ is left-computable, there exists a total recursive function $g : \mathbb{N}^+ \to \mathbb{Q}$ such that $g(m) \leq Z(T)$ for all $m \in \mathbb{N}^+$, and $\lim_{m\to\infty} g(m) = Z(T)$. We denote $Z(1)$ by β.

Given n and $\beta{\upharpoonright}{\lceil Tn \rceil}$, one can find a $k_0 \in \mathbb{N}^+$ such that

$$0.(\beta{\upharpoonright}{\lceil Tn \rceil}) < \sum_{i=1}^{k_0} 2^{-|p_i|}.$$

It is then easy to see that

$$\sum_{i=k_0+1}^{\infty} 2^{-|p_i|} < 2^{-Tn}.$$

Using the inequality $x^z + y^z \leq (x + y)^z$ for reals $x, y > 0$ and $z \geq 1$, it follows that

$$Z(T) - \widetilde{Z}_{k_0}(T) < 2^{-n}. \tag{5.6}$$

Note that $\widetilde{Z}_{k_0}(T) < \widetilde{Z}_{k_0}(f(l))$ for all $l \in \mathbb{N}^+$, and $\lim_{l \to \infty} \widetilde{Z}_{k_0}(f(l)) = \widetilde{Z}_{k_0}(T)$. Thus, since $\widetilde{Z}_{k_0}(T) < Z(T)$, one can then find $l_0, m_0 \in \mathbb{N}^+$ such that

$$\widetilde{Z}_{k_0}(f(l_0)) < g(m_0).$$

It follows from (5.6) and (5.5) that

$$2^{-n} > g(m_0) - \widetilde{Z}_{k_0}(T) > \widetilde{Z}_{k_0}(f(l_0)) - \widetilde{Z}_{k_0}(T) > 2^{-c}(f(l_0) - T).$$

Thus, $0 < f(l_0) - T < 2^{c-n}$. Note that $| f(l_0) - 0.(f(l_0) \restriction_n) | < 2^{-n}$ and moreover $| T - 0.(T \restriction_n) | < 2^{-n}$ due to (2.2). It follows that $| 0.(T \restriction_n) - 0.t_n | < (2^c + 2)2^{-n}$, where $f(l_0) \restriction_n$ is denoted by t_n. Hence $T \restriction_n = t_n, \, t_n \pm 1, \, t_n \pm 2, \, \ldots, \, t_n \pm (2^c + 1)$, where $T \restriction_n$ and t_n are regarded as a dyadic integer. Thus, one is left with $2^{c+1} + 3$ possibilities of $T \restriction_n$, so that one needs only $c + 2$ bits more in order to determine $T \restriction_n$.

Thus, there exists a partial recursive function $\Phi \colon \mathbb{N}^+ \times \{0, 1\}^* \times \{0, 1\}^* \to \{0, 1\}^*$ such that for every $n \in \mathbb{N}^+$ there exists $s \in \{0, 1\}^{c+2}$ with the property that $\Phi(n, \beta \restriction_{\lceil Tn \rceil}, s) = T \restriction_n$. Consider a prefix-free machine M such that, for every $p, v \in \{0, 1\}^*$, $M(p) = v$ if and only if there exist $q, r \in \operatorname{dom} U, t \in \{0, 1\}^*, s \in \{0, 1\}^{c+2}$, and $n \in \mathbb{N}^+$ with the properties that $p = qrts, U(q) = n, U(r) = |t|$, and $\Phi(n, t, s) = v$. Note that such a prefix-free machine M exists. Then, it is easy to see that

$$K_M(T \restriction_n) \leq K(n) + K(\lceil Tn \rceil) + |\beta \restriction_{\lceil Tn \rceil}| + c + 2$$

for every $n \in \mathbb{N}^+$. Thus, since $\lim_{n \to \infty} K(n)/n = 0$ and $\lim_{n \to \infty} K(\lceil Tn \rceil)/n = 0$ due to (2.7), and also $|\beta \restriction_{\lceil Tn \rceil}| \leq Tn + 1$ holds, it follows from (2.4) that T is T-compressible. $\qquad \square$

In a similar manner to the proof of Theorem 5.1, we can prove another version of a fixed point theorem on partial randomness as follows. Here, the weak Chaitin T-randomness is strengthen to the Chaitin T-randomness. Recall that $W(Q, T)$ is defined by (3.4).

Theorem 5.6 (Fixed Point Theorem on Partial Randomness II, Tadaki [44]) *Let Q be a computable real with $Q > 0$. For every $T \in (0, 1)$, if $W(Q, T)$ is a computable real, then the following hold:*

(i) T is right-computable and not left-computable.
(ii) T is Chaitin T-random and T-compressible. $\square$

5.2   Fixed Point Theorems by Thermodynamic Quantities of AIT

If we replace the role of $Z(T)$ in Theorem 5.1 with each of the three thermodynamic quantities $F(T)$, $E(T)$, and $S(T)$ of AIT in Definition 4.1, then the theorem still holds. Specifically, we can show the following three fixed point theorems on partial randomness (Tadaki [52]).[2]

Theorem 5.7 (Fixed Point Theorem by Helmholtz Free Energy) *For every* $T \in (0, 1)$, *if* $F(T)$ *is computable then* T *is weakly Chaitin* T-*random and* T-*compressible.*   □

Theorem 5.8 (Fixed Point Theorem by Energy) *For every* $T \in (0, 1)$, *if* $E(T)$ *is computable then* T *is Chaitin* T-*random and* T-*compressible.*   □

Theorem 5.9 (Fixed Point Theorem by Entropy) *For every* $T \in (0, 1)$, *if* $S(T)$ *is computable then* T *is Chaitin* T-*random and* T-*compressible.*   □

In Theorem 5.7, the computability of $Z(T)$ in the premise of Theorem 5.1 is replaced by the computability of $F(T)$. On the other hand, in Theorem 5.8 and Theorem 5.9, the computability of $W(Q, T)$ in the premise of Theorem 5.6 is replaced by the computability of $E(T)$ and $S(T)$, respectively. Thus, the weak Chaitin T-randomness of T in Theorems 5.1 is strengthen to the Chaitin T-randomness of T in Theorems 5.8 and 5.9.

The proof of Theorem 5.7 uses Theorems 5.11, 5.13, and 5.14 below. On the other hand, the proofs of Theorems 5.8 and 5.9 use Theorems 5.12, 5.13, and 5.14 below. All these proofs also use Theorem 5.10 below, where the thermodynamic relations in statistical mechanics are recovered by the thermodynamic quantities of AIT. We complete the proofs of Theorems 5.7, 5.8, and 5.9 in the next section. Compared with the proof of Theorem 5.7, the proofs of Theorems 5.8 and 5.9 are more delicate.

Recall that, in Definition 4.1, we have introduced the quantities $Z_k(T)$, $E_k(T)$, $F_k(T)$, and $S_k(T)$ based on the particular enumeration $q_1, q_2, q_3, q_4, \ldots$ of the set dom U. The enumeration $\{q_i\}$ have been chosen quite arbitrarily. Actually, as we saw in the preceding chapter, the values of the thermodynamic quantities $Z(T)$, $E(T)$, $F(T)$, and $S(T)$ for a real $T > 0$ do not depend on the choice of $\{q_i\}$, regardless of the convergence or divergence of them. Thus, in the rest of this chapter, we choose $\{q_i\}$ to be a *recursive* enumeration of dom U in Definition 4.1, for the sake of convenience. In particular, we denote $\{q_i\}$ by $\{p_i\}$ in order to emphasize the recursiveness of it.

Theorem 5.10 (Thermodynamic Relations)

(i) $F_k'(T) = -S_k(T)$, $E_k'(T) = C_k(T)$, and $S_k'(T) = C_k(T)/T$ *for every* $k \in \mathbb{N}^+$ *and every* $T \in (0, 1)$.

(ii) $F'(T) = -S(T)$, $E'(T) = C(T)$, and $S'(T) = C(T)/T$ *for every* $T \in (0, 1)$.

---

[2]It is still open whether a similar fixed point theorem holds for the specific heat $C(T)$ of AIT or not.

(iii) $S_k(T), C_k(T) \geq 0$ for every $k \in \mathbb{N}^+$ and every $T \in (0, 1)$. There exists $k_0 \in \mathbb{N}^+$ such that, for every $k \geq k_0$ and every $T \in (0, 1)$, it holds that $S_k(T), C_k(T) > 0$. Moreover, $S(T), C(T) > 0$ for every $T \in (0, 1)$.   □

The proof of Theorem 5.10 uses Lemma 5.1 below. For each $T \in (0, 1)$, we define $W(T)$ and $Y(T)$ as $\lim_{k \to \infty} W_k(T)$ and $\lim_{k \to \infty} Y_k(T)$, respectively, where $W_k(T) := \sum_{i=1}^{k} |p_i| 2^{-|p_i|/T}$ and $Y_k(T) := \sum_{i=1}^{k} |p_i|^2 2^{-|p_i|/T}$.

Lemma 5.1

(i)   For every $T \in (0, 1)$, it holds that $Z(T)$, $W(T)$, and $Y(T)$ converge and are positive reals.
(ii) The sequence $\{Z_k(T)\}_k$ of functions of T is uniformly convergent on $(0, 1)$ in the wider sense. The same holds for the sequences $\{W_k(T)\}_k$ and $\{Y_k(T)\}_k$.
(iii) The function $Z(T)$ of T is continuous on $(0, 1)$. The same holds for the functions $W(T)$ and $Y(T)$.

Proof (i) On the one hand, it follows immediately from Definition 4.1, that, for every $T \in (0, 1)$, it holds that $Z(T)$ converges and is a positive real. On the other hand, it follows from Theorem 3.4 (i) that, for every $T \in (0, 1)$, it holds that $W(T)$ and $Y(T)$ converge and are equal to the positive reals $W(1, T)$ and $W(2, T)$, respectively.

  (ii) Note that, for every $k \in \mathbb{N}^+$ and every $t, T \in (0, 1)$ with $t \leq T$,

$$0 < Z(t) - Z_k(t) = \sum_{i=k+1}^{\infty} 2^{-|p_i|/t} \leq \sum_{i=k+1}^{\infty} 2^{-|p_i|/T} = Z(T) - Z_k(T).$$

It follows that the sequence $\{Z_k(T)\}_k$ of functions of T is uniformly convergent on $(0, 1)$ in the wider sense. In the same manner, we can show that the sequences $\{W_k(T)\}_k$ and $\{Y_k(T)\}_k$ are uniformly convergent on $(0, 1)$ in the wider sense.

  (iii) Note that, for each $k \in \mathbb{N}^+$, the mapping $(0, 1) \ni T \mapsto Z_k(T)$ is a continuous function. It follows from Lemma 5.1 (ii) that the function $Z(T)$ of T is continuous on $(0, 1)$. In the same manner, we can show that the functions $W(T)$ and $Y(T)$ of T are continuous on $(0, 1)$.   □

The proof of Theorem 5.10 is then given as follows.

Proof (of Theorem 5.10) (i) First, from Definition 4.1, we see that, for every $k \in \mathbb{N}^+$ and every $T \in (0, 1)$,

$$F_k(T) = -T \log_2 Z_k(T), \tag{5.7}$$

$$E_k(T) = \frac{W_k(T)}{Z_k(T)}, \tag{5.8}$$

$$S_k(T) = \frac{W_k(T)}{T Z_k(T)} + \log_2 Z_k(T), \tag{5.9}$$

$$Z_k'(T) = \frac{\ln 2}{T^2} W_k(T), \tag{5.10}$$

$$W_k'(T) = \frac{\ln 2}{T^2} Y_k(T).$$

(5.11)

Thus, by straightforward differentiation, we can check that the relations of Theorem 5.10 (i) hold. For example, it follows from (5.9) and (5.8) that, for every $k \in \mathbb{N}^+$ and every $T \in (0, 1)$,

$$S_k'(T) = \frac{1}{T} E_k'(T) - \frac{1}{T^2} \frac{W_k(T)}{Z_k(T)} + \frac{1}{\ln 2} \frac{Z_k'(T)}{Z_k(T)}.$$

Using the definition $C_k(T) = E_k'(T)$ and the Eq. (5.10) we see that, for every $k \in \mathbb{N}^+$ and every $T \in (0, 1)$, it holds that $S_k'(T) = C_k(T)/T$.

(ii) From (5.8), (5.10), (5.11), and the definition $C_k(T) = E_k'(T)$, we see that, for every $k \in \mathbb{N}^+$ and every $T \in (0, 1)$,

$$C_k(T) = \frac{\ln 2}{T^2} \left\{ \frac{Y_k(T)}{Z_k(T)} - \left(\frac{W_k(T)}{Z_k(T)} \right)^2 \right\}.$$

(5.12)

First, recall from Theorem 4.2 (i) and Theorem 4.3 (i) that each of the limit values $F(T)$, $E(T)$, $S(T)$, and $C(T)$ exists as a real for every $T \in (0, 1)$. Using Lemma 5.1 in whole and the Eqs. (5.9) and (5.12), we can next check that the sequences $\{-S_k(T)\}_k$, $\{C_k(T)\}_k$ and $\{C_k(T)/T\}_k$ of functions of T are uniformly convergent on $(0, 1)$ in the wider sense. Hence, Theorem 5.10 (ii) follows immediately from Theorem 5.10 (i).

(iii) From (5.9) we see that, for every $k \in \mathbb{N}^+$ and every $T \in (0, 1)$,

$$S_k(T) = -\sum_{i=1}^{k} \frac{2^{-|p_i|/T}}{Z_k(T)} \log_2 \frac{2^{-|p_i|/T}}{Z_k(T)}.$$

Thus, $S_k(T) \geq 0$ for every $k \in \mathbb{N}^+$. We also see that, for every $k \geq 2$ and every $T \in (0, 1)$,

$$S_k(T) \geq -\frac{2^{-|p_1|/T}}{Z_k(T)} \log_2 \frac{2^{-|p_1|/T}}{Z_k(T)} > 0.$$

Hence, for every $T \in (0, 1)$,

$$S(T) \geq -\frac{2^{-|p_1|/T}}{Z(T)} \log_2 \frac{2^{-|p_1|/T}}{Z(T)} > 0.$$

On the other hand, from (5.12) we see that, for every $k \in \mathbb{N}^+$ and every $T \in (0, 1)$,

$$C_k(T) = \frac{\ln 2}{T^2} \sum_{i=1}^{k} \{|p_i| - E_k(T)\}^2 \frac{2^{-|p_i|/T}}{Z_k(T)}.$$

(5.13)

Thus, $C_k(T) \geq 0$ for every $k \in \mathbb{N}^+$ and every $T \in (0, 1)$. We note that there exists $l \in \mathbb{N}^+$ such that $|p_l| \leq |p_i|$ for every $i \in \mathbb{N}^+$. It is then easy to see that there exists $k_0 \in \mathbb{N}^+$ such that, for every $k \geq k_0$ and every $T \in (0, 1)$, it holds that $|p_l| < E_k(T)$. This is because there exists $i \in \mathbb{N}^+$ such that $|p_l| < |p_i|$. Thus, by (5.13) we see that, for every $k \geq \max\{l, k_0\}$ and every $T \in (0, 1)$,

$$C_k(T) \geq \frac{\ln 2}{T^2} \{|p_l| - E_k(T)\}^2 \frac{2^{-|p_l|/T}}{Z_k(T)} > 0. \tag{5.14}$$

It is also easy to see that $|p_l| < E(T)$ for every $T \in (0, 1)$. It follows from (5.14) that $C(T) > 0$ for every $T \in (0, 1)$. $\qquad\square$

Theorem 5.11 *Let $f : (0, 1) \to \mathbb{R}$. Suppose that f is increasing and there exists $g : (0, 1) \times \mathbb{N}^+ \to \mathbb{R}$ which satisfies the following four conditions:*

(i) *$\lim_{k \to \infty} g(T, k) = f(T)$ for every $T \in (0, 1)$.*

(ii) *$\{(q, r, k) \in \mathbb{Q} \times (\mathbb{Q} \cap (0, 1)) \times \mathbb{N}^+ \mid q < g(r, k)\}$ is an r.e. set.*

(iii) *For every $T \in (0, 1)$, there exist $a \in \mathbb{N}$, $k_0 \in \mathbb{N}^+$, and $t \in (T, 1)$ such that, for every $k \geq k_0$ and every $x \in (T, t)$, it holds that $g(x, k) - g(T, k) \leq 2^a(x - T)$.*

(iv) *For every $T \in (0, 1)$, there exist $b \in \mathbb{N}$ and $k_1 \in \mathbb{N}^+$ such that, for every $k \geq k_1$, it holds that $2^{-|p_{k+1}|/T-b} \leq g(T, k + 1) - g(T, k)$.*

Then, for every $T \in (0, 1)$, if $f(T)$ is right-computable then T is weakly Chaitin T-random.

Proof The proof of Theorem 5.11 is obtained by slightly simplifying the proof of Theorem 5.12 below. $\qquad\square$

Theorem 5.12 *Let $f : (0, 1) \to \mathbb{R}$. Suppose that f is increasing and there exists $g : (0, 1) \times \mathbb{N}^+ \to \mathbb{R}$ which satisfies the following four conditions:*

(i) *$\lim_{k \to \infty} g(T, k) = f(T)$ for every $T \in (0, 1)$.*

(ii) *$\{(q, r, k) \in \mathbb{Q} \times (\mathbb{Q} \cap (0, 1)) \times \mathbb{N}^+ \mid q < g(r, k)\}$ is an r.e. set.*

(iii) *For every $T \in (0, 1)$, there exist $a \in \mathbb{N}$, $k_0 \in \mathbb{N}^+$, and $t \in (T, 1)$ such that, for every $k \geq k_0$ and every $x \in (T, t)$, it holds that $g(x, k) - g(T, k) \leq 2^a(x - T)$.*

(iv) *For every $T \in (0, 1)$, there exist $b \in \mathbb{N}$, $c \in \mathbb{N}^+$, and $k_1 \in \mathbb{N}^+$ such that, for every $k \geq k_1$, it holds that $|p_{k+1}|^c 2^{-|p_{k+1}|/T-b} \leq g(T, k + 1) - g(T, k)$.*

Then, for every $T \in (0, 1)$, if $f(T)$ is right-computable then T is Chaitin T-random.

Proof Suppose that $T \in (0, 1)$ and $f(T)$ is right-computable. Then there exists a total recursive function $h : \mathbb{N}^+ \to \mathbb{Q}$ such that $f(T) \leq h(m)$ for all $m \in \mathbb{N}^+$ and $\lim_{m \to \infty} h(m) = f(T)$.

Since the condition (iii) holds for g, there exist $a \in \mathbb{N}$, $k_0 \in \mathbb{N}^+$, and $t \in (T, 1)$ such that

$$g(x, k) - g(T, k) \leq 2^a(x - T) \tag{5.15}$$

for every $k \geq k_0$ and every $x \in (T, t)$. We choose any particular $n_0 \in \mathbb{N}^+$ such that

$$0.(T{\restriction}_n) + 2^{-n} < t$$

for all $n \geq n_0$. Such n_0 exists since $T < t$ and $\lim_{n \to \infty} 0.(T{\restriction}_n) + 2^{-n} = T$. Due to (2.2) we further see that $T < 0.(T{\restriction}_n) + 2^{-n} < t$ for all $n \geq n_0$.

On the other hand, since the condition (iv) holds for g, there exist $b \in \mathbb{N}, c \in \mathbb{N}^+$, and $k_1 \in \mathbb{N}^+$ such that

$$|p_{k+1}|^c 2^{-|p_{k+1}|/T-b} \leq g(T, k+1) - g(T, k)$$

for every $k \geq k_1$. Without loss of generality, we can assume that $k_1 = k_0$. Thus, since $g(T, k)$ is increasing on k with $k \geq k_0$ and the condition (i) holds,

$$|p_i|^c 2^{-|p_i|/T-b} < f(T) - g(T, k) \tag{5.16}$$

if $i > k \geq k_0$.

Now, given $T{\restriction}_n$ with $n \geq n_0$, one can effectively find $k_e, m_e \in \mathbb{N}^+$ such that $k_e \geq k_0$ and $h(m_e) < g(0.(T{\restriction}_n) + 2^{-n}, k_e)$. This is possible because

$$f(T) < f(0.(T{\restriction}_n) + 2^{-n}),$$

$\lim_{k \to \infty} g(0.(T{\restriction}_n) + 2^{-n}, k) = f(0.(T{\restriction}_n) + 2^{-n})$, and the condition (ii) holds for g. It follows from $f(T) \leq h(m_e)$ and (5.15) that

$$f(T) - g(T, k_e) < g(0.(T{\restriction}_n) + 2^{-n}, k_e) - g(T, k_e) \leq 2^{a-n}.$$

It follows from (5.16) that, for every $i > k_e$, $|p_i|^c 2^{-|p_i|/T-b} < 2^{a-n}$ and therefore $cT \log_2 |p_i| - (a+b)T < |p_i| - Tn$. Thus, by calculating the set $\{ U(p_i) \mid i \leq k_e \}$ and picking any particular finite binary string which is not in this set, one can then obtain an $s \in \{0, 1\}^*$ such that $cT \log_2 K(s) - (a+b)T < K(s) - Tn$.

Hence, there exists a partial recursive function $\Psi : \{0, 1\}^* \to \{0, 1\}^*$ such that

$$cT \log_2 K(\Psi(T{\restriction}_n)) - (a+b)T < K(\Psi(T{\restriction}_n)) - Tn$$

for all $n \geq n_0$. Applying this inequality to itself, we have

$$cT \log_2 n < K(\Psi(T{\restriction}_n)) - Tn + O(1),$$

for all $n \geq n_0$. On the other hand, using (2.8), there is $d \in \mathbb{N}$ such that

$$K(\Psi(T{\restriction}_n)) \leq K(T{\restriction}_n) + d$$

for all $n \geq n_0$. It follows that $cT \log_2 n < K(T{\restriction}_n) - Tn + O(1)$. Hence, T is Chaitin T-random. $\qquad\square$

Theorem 5.13 *Let $f: (0, 1) \to \mathbb{R}$. Suppose that f is increasing and there exists $g: (0, 1) \times \mathbb{N}^+ \to \mathbb{R}$ which satisfies the following three conditions:*

(i)   For every $T \in (0, 1)$, $\lim_{k \to \infty} g(T, k) = f(T)$.
(ii) For every $T_1, T_2 \in (0, 1)$ with $T_1 < T_2$, there exists $k_0 \in \mathbb{N}^+$ such that, for every $k \geq k_0$ and every $x \in [T_1, T_2]$, it holds that $g(x, k) \leq f(x)$.
(iii) The set $\{(q, r, k) \in \mathbb{Q} \times (\mathbb{Q} \cap (0, 1)) \times \mathbb{N}^+ \mid q < g(r, k)\}$ is r.e.

Then, for every $T \in (0, 1)$, if $f(T)$ is right-computable then T is also right-computable.

Proof Suppose that $T \in (0, 1)$. We choose any particular $t_1, t_2 \in \mathbb{Q}$ with $0 < t_1 < T < t_2 < 1$. Then, since the condition (ii) holds for g, there exists $k_0 \in \mathbb{N}^+$ such that $g(x, k) \leq f(x)$ for every $k \geq k_0$ and every $x \in [t_1, t_2]$. Suppose further that $f(T)$ is right-computable. Then there exists a total recursive function $h: \mathbb{N}^+ \to \mathbb{Q}$ such that $f(T) \leq h(m)$ for all $m \in \mathbb{N}^+$ and $\lim_{m \to \infty} h(m) = f(T)$. Thus, since f is increasing and the condition (i) holds for g, we see that, for every $r \in \mathbb{Q} \cap [t_1, t_2]$, it holds that $T < r$ if and only if $\exists m \, \exists k \geq k_0 \, [h(m) < g(r, k)]$. Since the condition (iii) holds for g, the set $\{r \in \mathbb{Q} \cap [t_1, t_2] \mid \exists m \, \exists k \geq k_0 \, [h(m) < g(r, k)]\}$ is r.e. and therefore the set $\{r \in \mathbb{Q} \cap [t_1, t_2] \mid T < r\}$ is r.e. It follows from $T \in (t_1, t_2)$ that T is right-computable. $\square$

Theorem 5.14 *Let $f: (0, 1) \to \mathbb{R}$. Suppose that there exists $g: (0, 1) \times \mathbb{N}^+ \to \mathbb{R}$ which satisfies the following six conditions:*

(i)   For every $T \in (0, 1)$, $\lim_{k \to \infty} g(T, k) = f(T)$.
(ii) For every $T \in (0, 1)$, there exists $k_0 \in \mathbb{N}^+$ such that, for every $k \geq k_0$, it holds that $g(T, k) < f(T)$.
(iii) For every $T \in (0, 1)$, there exist $a \in \mathbb{N}$, $k_1 \in \mathbb{N}^+$, and $t \in (T, 1)$ such that, for every $k \geq k_1$ and every $x \in (T, t)$, it holds that $g(x, k) - g(T, k) \geq 2^{-a}(x - T)$.
(iv) For every $T \in (0, 1)$, there exist $b \in \mathbb{N}$, $c \in \mathbb{N}$, and $k_2 \in \mathbb{N}^+$ such that, for every $k \geq k_2$, it holds that $g(T, k+1) - g(T, k) \leq |p_{k+1}|^b \, 2^{-|p_{k+1}|/T+c}$.
(v)   For each $k \in \mathbb{N}^+$, the mapping $(0, 1) \ni T \mapsto g(T, k)$ is a continuous function.
(vi) $\{(q, r, k) \in \mathbb{Q} \times (\mathbb{Q} \cap (0, 1)) \times \mathbb{N}^+ \mid q > g(r, k)\}$ is an r.e. set.

Then, for every $T \in (0, 1)$, if $f(T)$ is left-computable and T is right-computable, then T is T-compressible.

Proof Suppose that $T \in (0, 1)$. Since the condition (ii) holds for g, there exists $k_0 \in \mathbb{N}^+$ such that

$$g(T, k) < f(T) \tag{5.17}$$

for every $k \geq k_0$. Since the condition (iii) holds for g, there exist $a \in \mathbb{N}$, $k_1 \in \mathbb{N}^+$, and $t \in (T, 1)$ such that

$$g(x, k) - g(T, k) \geq 2^{-a}(x - T) \tag{5.18}$$

for every $k \geq k_1$ and every $x \in (T, t)$. Since the condition (iv) holds for g, there exist $b \in \mathbb{N}$, $c \in \mathbb{N}$, and $k_2 \in \mathbb{N}^+$ such that

$$g(T, k+1) - g(T, k) \leq |p_{k+1}|^b \, 2^{-|p_{k+1}|/T+c} \tag{5.19}$$

for every $k \geq k_2$. Without loss of generality, we can assume that $k_0 = k_1 = k_2$.

Suppose further that T is right-computable and $f(T)$ is left-computable. Then there exists a total recursive function $A \colon \mathbb{N}^+ \to \mathbb{Q}$ such that $T < A(l) < t$ for all $l \in \mathbb{N}^+$ and $\lim_{l \to \infty} A(l) = T$, and there exists a total recursive function $B \colon \mathbb{N}^+ \to \mathbb{Q}$ such that $B(m) \leq f(T)$ for all $m \in \mathbb{N}^+$ and $\lim_{m \to \infty} B(m) = f(T)$.

Let u be an arbitrary computable real with $T < u < 1$. We denote $\sum_{i=1}^{\infty} |p_i|^b \, 2^{-|p_i|/u}$ by β. Note that $\beta = Z(u)$ if $b = 0$ and $\beta = W(b, u)$ otherwise. Thus, the infinite sum β converges and is weakly Chaitin u-random, due to Theorem 3.1 (i) or to Theorem 3.4 (i) depending on the value of $b \in \mathbb{N}$. It follows that the infinite binary sequence Binary (β) contains infinitely many ones, as well as infinitely many zeros.

Given n and $\beta \restriction_{\lceil Tn/u \rceil}$, one can effectively find a $k_e \in \mathbb{N}^+$ such that $k_e \geq k_0$ and

$$0.(\beta \restriction_{\lceil Tn/u \rceil}) + \lfloor \beta \rfloor < \sum_{i=1}^{k_e} |p_i|^b \, 2^{-|p_i|/u}.$$

This is possible since $0.(\beta \restriction_{\lceil Tn/u \rceil}) + \lfloor \beta \rfloor < \beta$ and $\lim_{k \to \infty} \sum_{i=1}^{k} |p_i|^b \, 2^{-|p_i|/u} = \beta$. Since $\beta - (0.(\beta \restriction_{\lceil Tn/u \rceil}) + \lfloor \beta \rfloor) \leq 2^{-\lceil Tn/u \rceil} \leq 2^{-Tn/u}$, it is then shown that

$$\sum_{i=k_e+1}^{\infty} |p_i|^b \, 2^{-|p_i|/u} = \beta - \sum_{i=1}^{k_e} |p_i|^b \, 2^{-|p_i|/u} < 2^{-Tn/u}.$$

Raising both ends of this inequality to the power u/T and using the inequality $x^z + y^z \leq (x+y)^z$ for reals $x, y > 0$ and $z \geq 1$, it is seen that

$$\sum_{i=k_e+1}^{\infty} |p_i|^b \, 2^{-|p_i|/T} \leq \sum_{i=k_e+1}^{\infty} |p_i|^{bu/T} \, 2^{-|p_i|/T} < 2^{-n}.$$

Using (5.19) and the condition (i), it follows that

$$f(T) - g(T, k_e) < \sum_{i=k_e+1}^{\infty} |p_i|^b \, 2^{-|p_i|/T+c} < 2^{c-n}. \tag{5.20}$$

On the other hand, $g(T, k_e) = \lim_{l \to \infty} g(A(l), k_e)$ since the condition (v) holds for g. Obviously, $g(T, k_e) < f(T)$ by (5.17). Thus, since the condition (vi) holds for g, one can then effectively find $l_e, m_e \in \mathbb{N}^+$ such that $g(A(l_e), k_e) < B(m_e)$. It follows from (5.20) and (5.18) that

$$2^{c-n} > f(T) - g(T, k_e) \geq B(m_e) - g(T, k_e) > g(A(l_e), k_e) - g(T, k_e) \geq 2^{-a}(A(l_e) - T).$$

Thus, $0 < A(l_e) - T < 2^{a+c-n}$. Note that $|A(l_e) - 0.(A(l_e)\restriction_n)| < 2^{-n}$ and moreover $|T - 0.(T\restriction_n)| < 2^{-n}$ due to (2.2). It follows that $|0.(T\restriction_n) - 0.r_n| < (2^{a+c} + 2)2^{-n}$, where $A(l_e)\restriction_n$ is denoted by r_n. Hence, $T\restriction_n = r_n,\ r_n \pm 1,\ r_n \pm 2,\ \ldots,\ r_n \pm (2^{a+c} + 1)$, where $T\restriction_n$ and r_n are regarded as a dyadic integer. Thus, one is left with $2^{a+c+1} + 3$ possibilities of $T\restriction_n$, so that one needs only $a + c + 3$ bits more in order to determine $T\restriction_n$.

Thus, there exists a partial recursive function $\Phi \colon \mathbb{N}^+ \times \{0, 1\}^* \times \{0, 1\}^* \to \{0, 1\}^*$ such that for every $n \in \mathbb{N}^+$ there exists $s \in \{0, 1\}^{a+c+3}$ with the property that $\Phi(n, \beta\restriction_{\lceil Tn/u \rceil}, s) = T\restriction_n$. Consider a prefix-free machine M such that, for every $p, v \in \{0, 1\}^*$, $M(p) = v$ if and only if there exist $q, r \in \mathrm{dom}\, U,\ t \in \{0, 1\}^*,\ s \in \{0, 1\}^{a+c+3}$, and $n \in \mathbb{N}^+$ with the properties that $p = qrts$, $U(q) = \dot{n}$, $U(r) = |t|$, and $\Phi(n, t, s) = v$. Note that such a prefix-free machine M exists. Then, it is easy to see that

$$K_M(T\restriction_n) \leq K(n) + K(\lceil Tn/u \rceil) + |\beta\restriction_{\lceil Tn/u \rceil}| + a + c + 3$$

for every $n \in \mathbb{N}^+$. Thus, since $\lim_{n \to \infty} K(n)/n = 0$ and $\lim_{n \to \infty} K(\lceil Tn/u \rceil)/n = 0$ due to (2.7), and also $|\beta\restriction_{\lceil Tn/u \rceil}| \leq Tn/u + 1$ holds, it follows from (2.4) that T is T/u-compressible. Since u is an arbitrary computable real with $T < u < 1$, it follows that T is T-compressible. □

5.3　The Proofs of the Fixed Point Theorems

In this section we complete the proofs of Theorems 5.7, 5.8, and 5.9.

5.3.1　The Proof of Theorem 5.7

We first complete the proof of Theorem 5.7, based on Theorems 5.10, 5.11, 5.13, and 5.14, as follows.

Let $f \colon (0, 1) \to \mathbb{R}$ with $f(T) = -F(T)$, and let $g \colon (0, 1) \times \mathbb{N}^+ \to \mathbb{R}$ with $g(T, k) = -F_k(T)$. First, it follows from Theorem 5.10 (ii) and (iii) that $-F'(T) = S(T) > 0$ for every $T \in (0, 1)$. Thus f is increasing.

Obviously, $\lim_{k \to \infty} g(T, k) = f(T)$ for every $T \in (0, 1)$. Using the mean value theorem we see that, for every $T \in (0, 1)$ and every $k \in \mathbb{N}^+$,

$$\frac{2^{-|p_{k+1}|/T}}{Z_{k+1}(T)} < \ln Z_{k+1}(T) - \ln Z_k(T) < \frac{2^{-|p_{k+1}|/T}}{Z_k(T)}. \tag{5.21}$$

It follows that, for every $T \in (0, 1)$ and every $k \in \mathbb{N}^+$, $g(T, k) < g(T, k + 1)$ and therefore $g(T, k) < f(T)$. At this point, the conditions (i) and (ii) of Theorem 5.11, all conditions of Theorem 5.13, and the conditions (i), (ii), (v), and (vi) of Theorem 5.14 hold for g.

Using (5.21) we see that, for every $T \in (0, 1)$ and every $k \in \mathbb{N}^+$,

$$\frac{T2^{-|p_{k+1}|/T}}{Z_{k+1}(T)\ln 2} < g(T, k + 1) - g(T, k) < \frac{T2^{-|p_{k+1}|/T}}{Z_k(T)\ln 2}.$$

Thus, the condition (iv) of Theorem 5.11 and the condition (iv) of Theorem 5.14 hold for g.

Using the mean value theorem and Theorem 5.10 (i) and (iii), we see that

$$S_k(T)(x - T) \leq g(x, k) - g(T, k) \leq S_k(t)(x - T) \tag{5.22}$$

for every $k \in \mathbb{N}^+$ and every $T, x, t \in (0, 1)$ with $T < x < t$. On the other hand, we see that, for every $k \in \mathbb{N}^+$ and every $T \in (0, 1)$,

$$E_{k+1}(T) - E_k(T) = \frac{Z_k(T)|p_{k+1}| - W_k(T)}{Z_{k+1}(T)Z_k(T)}2^{-|p_{k+1}|/T}.$$

Recall here that, for every $T \in (0, 1)$, each of $\lim_{k\to\infty} Z_k(T)$ and $\lim_{k\to\infty} W_k(T)$ exists as a positive real by Lemma 5.1 (i). It follows from $\lim_{k\to\infty} |p_{k+1}| = \infty$ that, for every $T \in (0, 1)$, there exists $k_0 \in \mathbb{N}^+$ such that, for every $k \geq k_0$, it holds that $E_k(T) < E_{k+1}(T)$ and therefore $S_k(T) < S_{k+1}(T)$ by Definition 4.1, and (5.21). Using Theorem 5.10 (iii), we see that, for every $T \in (0, 1)$, there exists $k_1 \in \mathbb{N}^+$ such that, for every $k \geq k_1$, it holds that $0 < S_{k_1}(T) \leq S_k(T) < S(T)$. Thus, using (5.22), for every $T, t \in (0, 1)$ with $T < t$, there exists $k_2 \in \mathbb{N}^+$ such that $S_{k_2}(T) > 0$ and for every $k \geq k_2$ and every $x \in (T, t)$,

$$S_{k_2}(T)(x - T) \leq g(x, k) - g(T, k) < S(t)(x - T).$$

Therefore, the condition (iii) of Theorem 5.11 and the condition (iii) of Theorem 5.14 hold for g.

Thus, Theorem 5.11, Theorem 5.13, and Theorem 5.14 result in the following three theorems, respectively.

Theorem 5.15 *For every $T \in (0, 1)$, if $F(T)$ is left-computable then T is weakly Chaitin T-random.* □

Theorem 5.16 *For every $T \in (0, 1)$, if $F(T)$ is left-computable then T is right-computable.* □

Theorem 5.17 *For every $T \in (0, 1)$, if both $F(T)$ and T are right-computable then T is T-compressible.* □

Theorem 5.7 follows immediately from these three theorems.

5.3.2   The Proof of Theorem 5.8

We complete the proof of Theorem 5.8, based on Theorems 5.10, 5.12, 5.13, and 5.14, as follows.

Let $f: (0, 1) \to \mathbb{R}$ with $f(T) = E(T)$, and let $g: (0, 1) \times \mathbb{N}^+ \to \mathbb{R}$ with $g(T, k) = E_k(T)$. First, it follows from Theorem 5.10 (ii) and (iii) that $E'(T) = C(T) > 0$ for every $T \in (0, 1)$. Thus f is increasing.

Obviously, $\lim_{k \to \infty} g(T, k) = f(T)$ for every $T \in (0, 1)$. At this point, the conditions (i) and (ii) of Theorem 5.12, the conditions (i) and (iii) of Theorem 5.13, and the conditions (i), (v), and (vi) of Theorem 5.14 hold for g.

We see that, for every $k \in \mathbb{N}^+$ and every $T \in (0, 1)$,

$$E_{k+1}(T) - E_k(T) = \frac{Z_k(T) |p_{k+1}| - W_k(T)}{Z_{k+1}(T) Z_k(T)} 2^{-|p_{k+1}|/T}. \qquad (5.23)$$

Recall here that, for every $T \in (0, 1)$, each of $\lim_{k \to \infty} Z_k(T)$ and $\lim_{k \to \infty} W_k(T)$ exists as a positive real by Lemma 5.1 (i). It follows from $\lim_{k \to \infty} |p_{k+1}| = \infty$ that, for every $T \in (0, 1)$, there exist $a \in \mathbb{N}$, $b \in \mathbb{N}$, and $k_0 \in \mathbb{N}^+$ such that, every $k \geq k_0$,

$$|p_{k+1}| 2^{-|p_{k+1}|/T-a} \leq g(T, k + 1) - g(T, k) \leq |p_{k+1}| 2^{-|p_{k+1}|/T+b}. \qquad (5.24)$$

Thus, the condition (iv) of Theorem 5.12 and the condition (iv) of Theorem 5.14 hold for g. It follows from (5.24) that, for every $T \in (0, 1)$, there exists $k_0 \in \mathbb{N}^+$ such that, every $k \geq k_0$, $g(T, k) < g(T, k + 1)$ and therefore $g(T, k) < f(T)$. Thus, the condition (ii) of Theorem 5.14 holds for g.

Using Lemma 5.1 (ii) and (iii), in addition to Lemma 5.1 (i), we can show a stronger statement than the inequalities (5.24). The stronger statement needed here is about the lower bound of (5.24). That is, based on (5.23), Lemma 5.1, and $\lim_{k \to \infty} |p_{k+1}| = \infty$, we can show that, for every $T_1, T_2 \in (0, 1)$ with $T_1 < T_2$, there exist $a \in \mathbb{N}$ and $k_0 \in \mathbb{N}^+$ such that, every $k \geq k_0$ and every $x \in [T_1, T_2]$,

$$|p_{k+1}| 2^{-|p_{k+1}|/x-a} \leq g(x, k + 1) - g(x, k).$$

It follows that the condition (ii) of Theorem 5.13 holds for g.

Now, using the mean value theorem and Theorem 5.10 (i), we see that, for every $k \in \mathbb{N}^+$ and every $T, x \in (0, 1)$ with $T < x$, there exists $y \in (T, x)$ such that

$$g(x, k) - g(T, k) = C_k(y)(x - T). \qquad (5.25)$$

On the other hand, using (5.12) we see that, for every $k \in \mathbb{N}^+$ and every $T \in (0, 1)$, $C_{k+1}(T) - C_k(T)$ is calculated as

$$\frac{\ln 2}{T^2} \frac{2^{-|p_{k+1}|/T}}{Z_{k+1}(T)} \left[|p_{k+1}|^2 - \left\{ \frac{W_{k+1}(T)}{Z_{k+1}(T)} + \frac{W_k(T)}{Z_k(T)} \right\} |p_{k+1}| \right.$$
$$\left. + \left\{ \frac{W_{k+1}(T)}{Z_{k+1}(T)} + \frac{W_k(T)}{Z_k(T)} \right\} \frac{W_k(T)}{Z_k(T)} - \frac{Y_k(T)}{Z_k(T)} \right].$$

Thus, based on Lemma 5.1 and $\lim_{k \to \infty} |p_{k+1}| = \infty$, we can show that, for every $T_1, T_2 \in (0, 1)$ with $T_1 < T_2$, there exist $a \in \mathbb{N}$ and $k_0 \in \mathbb{N}^+$ such that, every $k \geq k_0$ and every $y \in [T_1, T_2]$,

$$|p_{k+1}|^2 \, 2^{-|p_{k+1}|/y-a} \leq C_{k+1}(y) - C_k(y).$$

It follows from Theorem 5.10 (iii) that, for every $T_1, T_2 \in (0, 1)$ with $T_1 < T_2$, there exists $k_0 \in \mathbb{N}^+$ such that, every $k \geq k_0$ and every $y \in [T_1, T_2]$,

$$0 < \min C_{k_0}([T_1, T_2]) \leq C_k(y) < \max C([T_1, T_2]), \tag{5.26}$$

where

$$\min C_{k_0}([T_1, T_2]) := \min\{ C_{k_0}(z) \mid z \in [T_1, T_2] \},$$
$$\max C([T_1, T_2]) := \max\{ C(z) \mid z \in [T_1, T_2] \}.$$

In particular, $\max C([T_1, T_2])$ exists. This is because the function $C(T)$ of T is continuous on $(0, 1)$ by (5.12) and Lemma 5.1 (i) and (iii). It follows from (5.25) and (5.26) that, for every $T, t \in (0, 1)$ with $T < t$, there exist $a \in \mathbb{N}$, $b \in \mathbb{N}$, and $k_0 \in \mathbb{N}^+$ such that, for every $k \geq k_0$ and every $x \in (T, t)$,

$$2^{-a}(x - T) \leq g(x, k) - g(T, k) < 2^b(x - T).$$

Therefore, the condition (iii) of Theorem 5.12 and the condition (iii) of Theorem 5.14 hold for g.

Thus, Theorem 5.12, Theorem 5.13, and Theorem 5.14 result in the following three theorems, respectively.

Theorem 5.18 *For every $T \in (0, 1)$, if $E(T)$ is right-computable then T is Chaitin T-random.* □

Theorem 5.19 *For every $T \in (0, 1)$, if $E(T)$ is right-computable then T is also right-computable.* □

Theorem 5.20 *For every $T \in (0, 1)$, if $E(T)$ is left-computable and T is right-computable, then T is T-compressible.* □

Theorem 5.8 follows immediately from these three theorems.

5.3.3 The Proof of Theorem 5.9

In a similar manner to the proof of Theorem 5.8 described in the preceding sub-section, we can prove Theorem 5.9, based on Theorems 5.10, 5.12, 5.13, and 5.14. It is easy to convert the proof of Theorem 5.8 into the the proof of Theorem 5.9, because of the similarity between $E'_k(T) = C_k(T)$ and $S'_k(T) = C_k(T)/T$ given in Theorem 5.10 (i). In particular, Theorem 5.12, Theorem 5.13, and Theorem 5.14 result in the following three theorems, respectively, for the entropy $S(T)$ of AIT.

Theorem 5.21 *For every $T \in (0, 1)$, if $S(T)$ is right-computable then T is Chaitin T-random.*   □

Theorem 5.22 *For every $T \in (0, 1)$, if $S(T)$ is right-computable then T is also right-computable.*   □

Theorem 5.23 *For every $T \in (0, 1)$, if $S(T)$ is left-computable and T is right-computable, then T is T-compressible.*   □

Theorem 5.9 follows immediately from these three theorems.

5.3.4 Another Proof of Theorem 5.1 Based on the General Scheme

In Sect. 5.1 we have proved Theorem 5.1 in a *direct* manner. For completeness, we prove here Theorem 5.1, based on Theorems 5.11, 5.13, and 5.14, in a similar manner to the proof of Theorem 5.7 given in Sect. 5.3.1.

Let $f: (0, 1) \to \mathbb{R}$ with $f(T) = Z(T)$, and let $g: (0, 1) \times \mathbb{N}^+ \to \mathbb{R}$ with $g(T, k) = Z_k(T)$. First, it follows that f is increasing.

Obviously, $\lim_{k \to \infty} g(T, k) = f(T)$ for every $T \in (0, 1)$. It follows that, for every $T \in (0, 1)$ and every $k \in \mathbb{N}^+$,

$$g(T, k + 1) - g(T, k) = 2^{-|p_{k+1}|/T}.$$

Thus, for every $T \in (0, 1)$ and every $k \in \mathbb{N}^+$, $g(T, k) < g(T, k + 1)$ and therefore $g(T, k) < f(T)$. At this point, the conditions (i), (ii), and (iv) of Theorem 5.11, all conditions of Theorem 5.13, and the conditions (i), (ii), (iv), (v), and (vi) of Theorem 5.14 hold for g.

Using the mean value theorem we see that, for every $k \in \mathbb{N}^+$ and every $T, x, t \in (0, 1)$ with $T < x < t$,

$$\frac{\ln 2}{t^2} W_k(T)(x - T) < g(x, k) - g(T, k) < \frac{\ln 2}{T^2} W_k(t)(x - T),$$

where $W_k(T) = \sum_{i=1}^{k} |p_i| 2^{-|p_i|/T}$, as defined before Lemma 5.1 in Sect. 5.2. Note that, for every $t \in (0, 1)$, $W_k(t)$ is increasing on k, and $W(t) = \lim_{k \to \infty} W_k(t)$ exists as a real by Lemma 5.1 (i). Thus we see that

$$\frac{\ln 2}{t^2} W_1(T)(x - T) < g(x, k) - g(T, k) < \frac{\ln 2}{T^2} W(t)(x - T)$$

for every $k \in \mathbb{N}^+$ and every $T, x, t \in (0, 1)$ with $T < x < t$. Therefore, the condition (iii) of Theorem 5.11 and the condition (iii) of Theorem 5.14 hold for g.

Thus, Theorem 5.11, Theorem 5.13, and Theorem 5.14 result in Theorem 5.3, Theorem 5.4, and Theorem 5.5, respectively. Then, as we saw in Sect. 5.1, Theorem 5.1 follows immediately from the latter three theorems.

5.4 Some Properties of the Sufficient Conditions

In this section, we investigate some properties of the sufficient conditions for T to be a fixed point in the fixed point theorems on partial randomness; Theorems 5.1, 5.7, 5.8, and 5.9. We can show Theorems 5.24, 5.25, and 5.26 below for the sufficient conditions, in addition to Theorem 5.2 shown already for the sufficient condition regarding the partition function $Z(T)$ of AIT in Sect. 5.1.

Theorem 5.24 *The set $\{ T \in (0, 1) \mid F(T) \text{ is computable} \}$ is dense in $(0, 1)$.*

Proof It follows from Theorem 5.10 (ii) and (iii) that the function $F(T)$ of T is a decreasing continuous function on $(0, 1)$. Since the set of all computable reals is dense in $\mathbb{R}$, the result follows. □

In the same manner as the proof of Theorem 5.24, we can prove the following theorems for each of the energy $E(T)$ and the entropy $S(T)$ of AIT.

Theorem 5.25 *The set $\{ T \in (0, 1) \mid E(T) \text{ is computable} \}$ is dense in $(0, 1)$.* □

Theorem 5.26 *The set $\{ T \in (0, 1) \mid S(T) \text{ is computable} \}$ is dense in $(0, 1)$.* □

The following relation holds for the Helmholtz free energy $F(T)$ and the partition function $Z(T)$ of AIT, just as in normal statistical mechanics:

$$F(T) = -T \log_2 Z(T). \tag{5.27}$$

Based on this, we can show the following theorem regarding the relation between the sufficient conditions in the fixed point theorems by $Z(T)$ and $F(T)$.

Theorem 5.27 *There does not exist $T \in (0, 1)$ such that both $Z(T)$ and $F(T)$ are computable.*

Proof Contrarily, assume that both $Z(T)$ and $F(T)$ are computable for some $T \in (0, 1)$. Since $F(T) = -T \log_2 Z(T)$ and $0 < Z(T) < 1$, it is easy to see that T is computable. It follows from Theorem 4.2 (i) that $Z(T)$ is weakly Chaitin T-random. However, this contradicts the assumption that $Z(T)$ is computable, and the result follows. $\square$

Thus, the computability of $F(T)$ gives completely different fixed points from the computability of $Z(T)$. This implies that *neither the computability of $Z(T)$ nor the computability of $F(T)$ is the necessary condition for T to be a fixed point on partial randomness at all.* Actually, combining Theorem 5.27 with Theorem 5.24 and Theorem 5.7, it is easy to see that the set of all reals T such that the value $Z(T)$ is not computable but the partial randomness of T equals T is dense in $(0, 1)$.

In a similar manner, we can prove the following two theorems using the relations $S(T) = E(T)/T + \log_2 Z(T)$ and $S(T) = (E(T) - F(T))/T$, respectively.

Theorem 5.28 *There does not exist $T \in (0, 1)$ such that $Z(T)$, $E(T)$, and $S(T)$ are all computable.* $\square$

Theorem 5.29 *There does not exist $T \in (0, 1)$ such that $F(T)$, $E(T)$, and $S(T)$ are all computable.* $\square$

Using the property of a fixed point in the fixed point theorems on partial randomness, we can show the following theorem.

Theorem 5.30 $S_a \cap S_b = \emptyset$ *for any distinct computable reals $a, b \in (0, 1]$, where $S_a := \{ T \in (0, 1) \mid Z(aT) \text{ is computable} \}$.*

Proof Let $T \in (0, 1)$, and let a be a computable real with $a \in (0, 1]$. Suppose that $Z(aT)$ is computable. It follows from Theorem 5.1 that $\lim_{n\to\infty} K((aT)\restriction_n)/n = aT$. Since a is computable, note that $K((aT)\restriction_n) = K(T\restriction_n) + O(1)$ for all $n \in \mathbb{N}^+$. It follows that $\lim_{n\to\infty} K(T\restriction_n)/n = aT$.

Thus, for every computable reals $a, b \in (0, 1]$, if $S_a \cap S_b \neq \emptyset$ then $a = b$. This completes the proof. $\square$

As a corollary of Theorem 5.30, we have the following, for example.

Corollary 5.2 *For every $T \in (0, 1)$, if $Z(T)$ is computable, then $Z(T/n)$ is not computable for every $n \in \mathbb{N}^+$ with $n \geq 2$. In other words, for every $T \in (0, 1)$, if the sum $\sum_{i=1}^{\infty} 2^{-|p_i|/T}$ is computable, then the corresponding power sum $\sum_{i=1}^{\infty} \left(2^{-|p_i|/T}\right)^n$ is not computable for every $n \in \mathbb{N}^+$ with $n \geq 2$.* $\square$

5.5 Thermodynamic Quantities Based on an Arbitrary Optimal Prefix-Free Machine

The thermodynamic quantities of AIT are defined based on the domain dom U of the definition of the optimal prefix-free machine U, as we did so in Definition 4.1. If we choose an arbitrary optimal prefix-free machine V in the definition of the thermodynamic quantities of AIT, we can define another thermodynamic quantities of AIT based on dom V, instead of dom U. Thus, we can define the *partition function* $Z_V(T)$, the *energy* $E_V(T)$, the *Helmholtz free energy* $F_V(T)$, the *entropy* $S_V(T)$, and the *specific heat* $C_V(T)$ in the same manner as in Definition 4.1, but using an arbitrary optimal prefix-free machine V instead of U. Then, obviously, the following theorems of the same form as Theorems 4.2 and 4.3 hold for an arbitrary optimal prefix-free machine V instead of U.

Theorem 5.31 *Let V be an optimal prefix-free machine, and let $T \in \mathbb{R}$.*

(i) *If $0 < T \leq 1$ and T is computable, then each of $Z_V(T)$ and $F_V(T)$ converges and is weakly Chaitin T-random and T-compressible.*

(ii) *If $1 < T$, then $Z_V(T)$ and $F_V(T)$ diverge to ∞ and $-\infty$, respectively.* □

Theorem 5.32 *Let V be an optimal prefix-free machine, and let $T \in \mathbb{R}$.*

(i) *If $0 < T < 1$ and T is computable, then each of $E_V(T)$, $S_V(T)$, and $C_V(T)$ converges and is Chaitin T-random and T-compressible.*

(ii) *If $T = 1$, then all of $E_V(T)$, $S_V(T)$, and $C_V(T)$ diverge to ∞.*

(iii) *If $1 < T$, then both $E_V(T)$ and $S_V(T)$ diverge to ∞.* □

We denote by $\mathscr{FP}_w$ the set of all real $T \in (0, 1)$ such that T is weakly Chaitin T-random and T-compressible, and denote by $\mathscr{FP}$ the set of all real $T \in (0, 1)$ such that T is Chaitin T-random and T-compressible. Obviously, $\mathscr{FP} \subset \mathscr{FP}_w$. Each element T of $\mathscr{FP}_w$ is a fixed point on partial randomness.

Let V be a prefix-free machine. We define a set $\mathscr{Z}(V)$ by

$$\mathscr{Z}(V) := \{\, T \in (0, 1) \mid Z_V(T) \text{ is computable} \,\}.$$

In the same manner, we define sets $\mathscr{F}(V)$, $\mathscr{E}(V)$, and $\mathscr{S}(V)$ based on the computability of $F_V(T)$, $E_V(T)$, and $S_V(T)$, respectively. Since V is an optimal prefix-free machine, fixed point theorems on partial randomness of the same form as Theorems 5.1, 5.7, 5.8, and 5.9 hold for V instead of U. In other words, we have the following.

Theorem 5.33 *Let V be an optimal prefix-free machine. Then $\mathscr{Z}(V) \cup \mathscr{F}(V) \subset \mathscr{FP}_w$ and $\mathscr{E}(V) \cup \mathscr{S}(V) \subset \mathscr{FP}$.* □

Chapter 6
Statistical Mechanical Meaning of the Thermodynamic Quantities of AIT

6.1 Perfect Correspondence to Normal Statistical Mechanics

Generally speaking, in order to give a statistical mechanical interpretation to a framework which looks unrelated to statistical mechanics at first glance, it is important to identify a *microcanonical ensemble* in the framework. Once we can do so, we can easily develop an equilibrium statistical mechanics on the framework according to the theoretical development of normal equilibrium statistical mechanics. Recall here that the microcanonical ensemble is a certain sort of uniform probability distribution (see Sect. 1.3). In fact, in Chap. 1 we have developed a statistical mechanical interpretation of the noiseless source coding scheme in information theory by identifying a microcanonical ensemble in the scheme. Then, based on this identification, in Chap. 1 the notions in statistical mechanics such as statistical mechanical entropy, temperature, and thermal equilibrium are translated into the context of noiseless source coding.

Hence, in order to develop a total statistical mechanical interpretation of AIT, it is appropriate to identify a microcanonical ensemble in the framework of AIT. Note, however, that AIT is not a physical theory but a purely mathematical theory. Therefore, in order to obtain significant results for the development of AIT itself, we have to develop a statistical mechanical interpretation of AIT in a mathematically rigorous manner, unlike in normal statistical mechanics in physics where arguments are not necessarily mathematically rigorous. A fully rigorous mathematical treatment of statistical mechanics is already developed (see Ruelle [30]). At present, however, it would not as yet seem to be an easy task to merge AIT with this mathematical treatment in a satisfactory manner.

In Chaps. 4 and 5, for mathematical strictness we have developed a statistical mechanical interpretation of AIT in a different way from the idealism above. In

---

This chapter is an extended version of Tadaki [48].

Chap. 4 we have introduced the notion of *thermodynamic quantities at temperature T*, such as partition function $Z(T)$, energy $E(T)$, Helmholtz free energy $F(T)$, statistical mechanical entropy $S(T)$, and specific heat $C(T)$, into AIT by performing Replacements 4.1 for the corresponding thermodynamic quantities at temperature T in statistical mechanics.

In this chapter, according to Tadaki [48], we show that, if we do not stick to the mathematical strictness of an argument, we can certainly develop a *total statistical mechanical interpretation of AIT* which attains a *perfect correspondence to normal statistical mechanics*. In the total statistical mechanical interpretation, we identify a microcanonical ensemble in AIT in a similar manner to Chap. 1, based on the probability measure which gives Chaitin's Ω the meaning of the *halting probability* actually. This identification enables us to clarify the statistical mechanical meaning of the thermodynamic quantities of AIT, which are originally introduced in Definition 4.1 in a rigorous manner.

6.2 Total Statistical Mechanical Interpretation of AIT

In what follows, based on a physical argument we develop the total statistical mechanical interpretation of AIT realizing a perfect correspondence to normal equilibrium statistical mechanics. In consequence, we justify the interpretation of $\Omega(D)$ as a partition function and clarify the statistical mechanical meaning of the thermodynamic quantities of AIT introduced in Definition 4.1. In Chap. 1 we have developed a statistical mechanical interpretation of the noiseless source coding scheme based on an absolutely optimal instantaneous code by identifying a microcanonical ensemble in the scheme. In a similar manner to Chap. 1, we develop a total statistical mechanical interpretation of AIT. This can be possible because the set dom U is prefix-free and therefore the action of the optimal prefix-free machine U can be regarded as an instantaneous code which is extended over an infinite set. Note again that, in what follows, we do not stick to the mathematical strictness of the argument and we make an argument on the same level of mathematical strictness as statistical mechanics in physics. For the review of the basic framework of equilibrium statistical mechanics and its theoretical development, see Sects. 1.3 and 4.2.

Now we give a total statistical mechanical interpretation to AIT. As considered in Chaitin [9], think of the optimal prefix-free machine U as a decoding equipment at the receiving end of a noiseless binary communication channel. Regard its programs (i.e., finite binary strings in dom U) as codewords and regard the result of the computation by U, which is a finite binary string, as a decoded "symbol." Since dom U is an infinite prefix-free set, such codewords are thought to form an "instantaneous code" which is extended over an infinite set. Thus, successive symbols sent through the channel in the form of concatenation of codewords can be separated.

For establishing the total statistical mechanical interpretation of AIT, we assume that the infinite binary sequence sent through the channel is generated by infinitely repeated tosses of a fair coin. Under this assumption, the infinite binary sequence is referred to as *the channel infinite sequence* hereafter. For each $r \in \{0, 1\}^*$, let $Q(r)$ be the probability that the channel infinite sequence has the prefix r. It follows that $Q(r) = 2^{-|r|}$. Thus, the channel infinite sequence is the random variable drawn according to Lebesgue measure on $\{0, 1\}^\infty$. Note that the success probability for U to decode one symbol at the receiving end of the channel infinite sequence equals to Chaitin's halting probability Ω since $\Omega = \sum_{p \in \operatorname{dom} U} Q(p)$. Thus, since this success probability can be regarded as a probability for U to halt at the receiving end of the channel infinite sequence, we will establish the total statistical mechanical interpretation of AIT, based on a probability measure which gives Chaitin's Ω the meaning of *halting probability* actually. In addition, note that the probability to get a finite binary string s as the first decoded symbol by U at the receiving end of the channel infinite sequence equals to $P_U(s)$, which is a *universal probability* introduced in Sect. 2.3, since $P_U(s) = \sum_{U(p)=s} Q(p)$.

Let N be a large number, say $N \sim 10^{22}$. We relate AIT to the equilibrium statistical mechanics reviewed in Sect. 1.3, in the following manner. Among all infinite binary sequences, consider infinite binary sequences of the form $p_1 p_2 \cdots p_N \alpha$ with $p_1, p_2, \ldots, p_N \in \operatorname{dom} U$ and $\alpha \in \{0, 1\}^\infty$. For each i, the ith slot fed by p_i corresponds to the ith quantum subsystem $\mathscr{S}_i$. On the other hand, the ordered sequence of the 1st slot, the 2nd slot, $\ldots$, and the Nth slot corresponds to the quantum system $\mathscr{S}_{\text{total}}$. We relate a codeword $p \in \operatorname{dom} U$ to an energy eigenstate of a quantum subsystem specified by n, and relate the length $|p|$ of the codeword p to the energy E_n of the energy eigenstate of the quantum subsystem specified by n. Then, a finite binary string $p_1 \cdots p_N$ corresponds to an energy eigenstate of $\mathscr{S}_{\text{total}}$ specified by $(n_1, \ldots, n_N)$, and its length $|p_1 \cdots p_N|$, which equals to $|p_1| + \cdots + |p_N|$, corresponds to the energy $E_{n_1} + \cdots + E_{n_N}$ of the energy eigenstate of $\mathscr{S}_{\text{total}}$ specified by $(n_1, \ldots, n_N)$.

Generally speaking, in the statistical mechanical interpretation of AIT, a prefix-free machine corresponds to a quantum system. To be specific, the domain of the definition of a prefix-free machine M corresponds to the complete set of energy eigenstates of a quantum system $\mathscr{Q}$, where each program $p \in \operatorname{dom} M$ of M corresponds to an energy eigenstate of $\mathscr{Q}$. Then, for every program $p \in \operatorname{dom} M$, the length $|p|$ of p corresponds to the energy of the energy eigenstate to which the program p corresponds.

On that premise, let us consider the "composition" of two prefix-free machines M_1 and M_2. Suppose that M_1 and M_2 correspond to quantum systems $\mathscr{Q}_1$ and $\mathscr{Q}_2$ with state spaces $\mathscr{H}_1$ and $\mathscr{H}_2$, respectively. Then the composition of the prefix-free machines M_1 and M_2 corresponds to the composition of the quantum systems $\mathscr{Q}_1$ and $\mathscr{Q}_2$ in the following manner: Consider programs $p_1 \in \operatorname{dom} M_1$ and $p_2 \in \operatorname{dom} M_2$ which correspond to state vectors $\Psi_1 \in \mathscr{H}_1$ and $\Psi_2 \in \mathscr{H}_2$, respectively. Then the concatenation $p_1 p_2$ corresponds to the quantum state of the composite quantum system, which is represented by the tensor product $\Psi_1 \otimes \Psi_2 \in \mathscr{H}_1 \otimes \mathscr{H}_2$ of Ψ_1 and Ψ_2. Note here that $\mathscr{H}_1 \otimes \mathscr{H}_2$ is the state space of the composite quantum system consisting of $\mathscr{Q}_1$ and $\mathscr{Q}_2$. Put mathematically, *in the statistical mechanical interpretation of AIT the concatenation of finite binary strings corresponds to the tensor product of elements of Hilbert spaces*. Recall that the equality $|p_1 p_2| = |p_1| + |p_2|$ holds. This equality can be interpreted in the way that the "energy" of $p_1 p_2$ equals to the sum of the "energy" of p_1 and the "energy" of p_2. This is consistent with the *additivity of energy* in quantum mechanics.

On the one hand, note that, given a concatenation $p_1 p_2$, we can decompose it into p_1 and p_2 *uniquely* since dom M_1 is prefix-free. On the other hand, in quantum mechanics, given a quantum state of a composite quantum system, the quantum state of each subsystem is uniquely determined by taking the *partial trace* over the remaining subsystems. The unique decomposability of the concatenation of programs corresponds to this fact in quantum mechanics.

Now, we define a subset $C(L, N)$ of $\{0, 1\}^*$ as the set of all finite binary strings of the form $p_1 \cdots p_N$ with $p_i \in$ dom U whose total length $|p_1 \cdots p_N|$ lies between L and $L + \delta L$. Then, $\Theta(L, N)$ is defined as the total number of elements of the finite set $C(L, N)$. Therefore, $\Theta(L, N)$ is the total number of all concatenations of N codewords whose total length lies between L and $L + \delta L$. It follows that if $p_1 \cdots p_N \in C(L, N)$, then $2^{-(L+\delta L)} \leq Q(p_1 \cdots p_N) \leq 2^{-L}$. Thus, all concatenations $p_1 \cdots p_N \in C(L, N)$ of N codewords occur in a prefix of the channel infinite sequence with the same probability 2^{-L}. Note here that we care nothing about the magnitude of δL, as in the case of statistical mechanics. Thus, the following principle, called the *principle of equal conditional probability*, holds.

The Principle of Equal Conditional Probability: Given that a concatenation of N codewords of total length L occurs in a prefix of the channel infinite sequence, all such concatenations occur with the same probability $1/\Theta(L, N)$. □

We introduce a *microcanonical ensemble* into AIT in this manner. Thus, we can develop a certain sort of equilibrium statistical mechanics on AIT. This is possible because, in order to do so, we only have to follow the theoretical development of normal equilibrium statistical mechanics, starting from this microcanonical ensemble. Note that, in statistical mechanics, the principle of equal probability is just a conjecture which is not yet proved completely in a realistic physical system. In contrast, in our total statistical mechanical interpretation of AIT, the principle of equal conditional probability is automatically satisfied.

The *statistical mechanical entropy* $S(L, N)$ is defined by

$$S(L, N) := \log_2 \Theta(L, N). \tag{6.1}$$

The *temperature* $T(L, N)$ is then defined by

$$\frac{1}{T(L, N)} := \frac{\partial S}{\partial L}(L, N). \tag{6.2}$$

Thus, the temperature is a function of L and N.

According to the theoretical development of equilibrium statistical mechanics,[1] we can introduce a *canonical ensemble* into AIT in the following manner. We investigate the probability distribution of the left-most codeword p_1 of the channel infinite sequence, given that a concatenation of N codewords of total length L occurs in a prefix of the channel infinite sequence. For each $p \in \text{dom } U$, let $R(p)$ be the probability that the left-most codeword of the channel infinite sequence is p, given that a concatenation of N codewords of total length L occurs in a prefix of the channel infinite sequence. Then, based on the principle of equal conditional probability, we see that

$$R(p) = \Theta(L - |p|, N - 1)/\Theta(L, N).$$

From the general definition (6.1) of statistical mechanical entropy, we thus have

$$R(p) = 2^{S(L-|p|, N-1)-S(L,N)}. \tag{6.3}$$

Let $E(L, N)$ be the expected length of the left-most codeword of the channel infinite sequence, given that a concatenation of N codewords of total length L occurs in a prefix of the channel infinite sequence. Then, since N and L are extremely large, from the definition (6.1) of statistical mechanical entropy the following equality is expected to hold:

$$S(L, N) = S(E(L, N), 1) + S(L - E(L, N), N - 1). \tag{6.4}$$

Here, the term $S(L, N)$ in the left-hand side denotes the statistical mechanical entropy of the whole concatenation of N codewords of total length L. On the other hand, the first term $S(E(L, N), 1)$ in the right-hand side denotes the statistical mechanical entropy of the left-most codeword of the concatenation of N codewords of total length L while the second term $S(L - E(L, N), N - 1)$ in the right-hand side denotes the statistical mechanical entropy of the remaining $N - 1$ codewords of the concatenation of N codewords of total length L. Thus, the equality (6.4) represents the additivity of the statistical mechanical entropy. We assume here that the equality (6.4) holds, as we do so in normal equilibrium statistical mechanics.

By expanding $S(L - |p|, N - 1)$ around the "equilibrium point" $L - E(L, N)$, we have

$$S(L - |p|, N - 1) = S(L - E(L, N) + E(L, N) - |p|, N - 1)$$

$$= S(L - E(L, N), N - 1) + \frac{\partial S}{\partial L}(L - E(L, N), N - 1)(E(L, N) - |p|). \tag{6.5}$$

Here, we ignore the higher order terms than the first order. Since $N \gg 1$ and $L \gg E(L, N)$, using the definition (6.2) of temperature we have

---

[1]We follow the argument of Sect. 16-1 of Callen [2], in particular.

$$\frac{\partial S}{\partial L}(L - E(L, N), N - 1) = \frac{\partial S}{\partial L}(L, N) = \frac{1}{T(L, N)}. \tag{6.6}$$

Hence, by (6.5) and (6.6), we have

$$S(L - |p|, N - 1) = S(L - E(L, N), N - 1) + \frac{1}{T(L, N)}(E(L, N) - |p|). \tag{6.7}$$

Thus, using (6.3), (6.4), and (6.7), we obtain

$$R(p) = 2^{\frac{E(L,N) - T(L,N)S(E(L,N),1)}{T(L,N)}} 2^{-\frac{|p|}{T(L,N)}}. \tag{6.8}$$

Then, according to statistical mechanics or thermodynamics, we define the *Helmholtz free energy* $F(L, N)$ of the left-most codeword of the concatenation of N codewords of total length L by

$$F(L, N) := E(L, N) - T(L, N)S(E(L, N), 1). \tag{6.9}$$

It follows from (6.8) that

$$R(p) = 2^{\frac{F(L,N)}{T(L,N)}} 2^{-\frac{|p|}{T(L,N)}}. \tag{6.10}$$

Then, using $\sum_{p \in \text{dom } U} R(p) = 1$ we see that

$$1 = 2^{\frac{F(L,N)}{T(L,N)}} \sum_{p \in \text{dom } U} 2^{-\frac{|p|}{T(L,N)}}. \tag{6.11}$$

Thus, it follows from (6.10) that, for any $p \in \text{dom } U$,

$$R(p) = \frac{1}{Z(T(L, N))} 2^{-\frac{|p|}{T(L,N)}}, \tag{6.12}$$

where $Z(T)$ is defined by

$$Z(T) := \sum_{p \in \text{dom } U} 2^{-\frac{|p|}{T}} \quad (T > 0). \tag{6.13}$$

$Z(T)$ is called the *partition function* (of the left-most codeword of the channel infinite sequence). Thus, in our total statistical mechanical interpretation of AIT, the partition function $Z(T)$ has exactly the same form as $\Omega(D)$, which is defined by (3.1). The distribution in the form of $R(p)$ is called a *canonical ensemble* in statistical mechanics.

Then, using (6.11) and (6.13), $F(L, N)$ is calculated as

$$F(L, N) = F(T(L, N)), \tag{6.14}$$

where $F(T)$ is defined by

$$F(T) := -T \log_2 Z(T) \quad (T > 0). \tag{6.15}$$

On the other hand, from the definition of $R(p)$, $E(L, N)$ is calculated as

$$E(L, N) = \sum_{p \in \text{dom } U} |p| R(p).$$

Thus, using (6.12) we have

$$E(L, N) = E(T(L, N)), \tag{6.16}$$

where $E(T)$ is defined by

$$E(T) := \frac{1}{Z(T)} \sum_{p \in \text{dom } U} |p| 2^{-\frac{|p|}{T}} \quad (T > 0). \tag{6.17}$$

Then, using (6.9), (6.14), and (6.16), the statistical mechanical entropy $S(E(L, N), 1)$ of the left-most codeword of the concatenation of N codewords of total length L is calculated as $S(E(L, N), 1) = S(T(L, N))$, where $S(T)$ is defined by

$$S(T) := \{E(T) - F(T)\}/T \quad (T > 0). \tag{6.18}$$

Note that the statistical mechanical entropy $S(E(L, N), 1)$ coincides with the *Shannon entropy*

$$- \sum_{p \in \text{dom } U} R(p) \log_2 R(p)$$

of the distribution $R(p)$.

Finally, the *specific heat* $C(T)$ of the left-most codeword of the channel infinite sequence is defined by

$$C(T) := E'(T) \quad (T > 0), \tag{6.19}$$

where $E'(T)$ is the derived function of $E(T)$.

Thus, a total statistical mechanical interpretation of AIT has been established, based on a physical argument. We can check that the formulas in this argument: the partition function (6.13), the expected length of the left-most codeword (6.17), the Helmholtz free energy (6.15), the statistical mechanical entropy (6.18), and the specific heat (6.19) correspond to the thermodynamic quantities of AIT in Definition 4.1. Thus, the statistical mechanical meaning of the notion of thermodynamic quantities of AIT in Definition 4.1 is clarified by this argument.

6.3   Future Direction

In this chapter, we have developed a total statistical mechanical interpretation of AIT which actualizes a perfect correspondence to normal statistical mechanics. However, the argument used in the development is on the same level of mathematical strictness as normal statistical mechanics in physics. Thus, we try to make the argument a rigorous form in a future study. This effort might stimulate a further unexpected development of the mathematical research of AIT.

Chapter 7
The Partial Randomness of Recursively Enumerable Reals

7.1 R.E. Reals and Thermodynamic Quantities of AIT

In this chapter, we investigate the partial randomness of a *recursively enumerable real*, also known as a *left-computable real*, and gives a theorem which plays an important role in the analysis of the thermodynamic quantities of AIT.

As we introduced in Sect. 2.2, a real α is called *recursively enumerable* (*r.e.*, for short) if there exists a computable, increasing sequence of rationals which converges to α. In the statistical mechanical interpretation of AIT, it is important to study the properties of r.e. reals. This is because all the thermodynamic quantities of AIT are r.e. reals, as shown in the following theorem.

Theorem 7.1 *Let V be an optimal prefix-free machine.*

(i) *If T is a computable real with $0 < T < 1$, then each of $Z_V(T)$, $-F_V(T)$, $E_V(T)$, $S_V(T)$, and $C_V(T)$ is an r.e. real.*

(ii) *For every T with $0 < T < 1$, if one of the values $Z_V(T)$, $F_V(T)$, $E_V(T)$, and $S_V(T)$ is computable, then $-T$ is an r.e. real.*[1]

Proof Theorem 7.1 (i) with the optimal prefix-free machine U as V follows immediately from Theorem 4.4, Theorem 4.5 (i), Theorem 4.6 (i), Theorem 4.7 (i), and Theorem 4.8 (i). On the other hand, Theorem 7.1 (ii) with the optimal prefix-free machine U as V follows immediately from Theorem 5.4, Theorem 5.16, Theorem 5.19, and Theorem 5.22. Obviously, Theorem 7.1 (i) and (ii) hold for an arbitrary optimal prefix-free machine V instead of U. This is because U is chosen quite arbitrarily in Sect. 2.3. $\qquad\square$

The main result of this chapter is Theorem 7.5 below, which gives many equivalent characterizations of partial randomness for an r.e. real. Theorem 7.5 plays

---

[1] This real T is a fixed point on partial randomness as we saw in Chap. 5.

This chapter is a rearrangement of Tadaki [47].

© The Author(s), under exclusive license to Springer Nature Singapore Pte Ltd. 2019
K. Tadaki, *A Statistical Mechanical Interpretation of Algorithmic Information Theory*,
SpringerBriefs in Mathematical Physics 36,
https://doi.org/10.1007/978-981-15-0739-7_7

an important role in the analysis of the thermodynamic quantities of AIT, as inspired by Theorem 7.1.

The *randomness* of an r.e. real α can be characterized in various ways using each of the notions; *program-size complexity, Martin-Löf test, Chaitin Ω number,* the *domination* and Ω-*likeness* of α, the *universality* of a computable, increasing sequence of rationals which converges to α, and *universal probability.* These equivalent characterizations of randomness of an r.e. real are summarized in Theorem 7.3 (see Sect. 7.2), where the equivalences are established by a series of works of Martin-Löf [24], Schnorr [33], Chaitin [9], Solovay [37], Calude et al. [4], Kučera and Slaman [19], and Tadaki [42], between 1966 and 2006.

In this chapter, we generalize these characterizations of randomness over the notion of *partial randomness*. We introduce many characterizations of partial randomness for an r.e. real by parameterizing each of the notions above on randomness with a real $T \in (0, 1]$. In particular, we introduce the notion of T-*convergence* for a computable, increasing sequence of rationals and then introduce the same notion for an r.e. real. The notion of T-convergence plays a crucial role in these our characterizations of partial randomness for an r.e. real. We then prove the equivalence of *all* these characterizations of partial randomness in Theorem 7.5, the main result of this chapter, in Sect. 7.3.

As we referred to in Sect. 3.3, by a series of works of Ryabko [31, 32], Staiger [38, 39], Tadaki [40, 41], Lutz [23], and Mayordomo [25] over two decades, the equivalence between the notion of compression rate by program-size complexity (or the notion of partial randomness) and Hausdorff dimension seems to be established at present. The subject of the equivalence is one of the most active areas of the recent research of AIT. In the context of the subject, we can consider the notion of the *dimension* of an individual real in particular, and this notion plays a crucial role in the subject. As one of the main applications of the main result, Theorem 7.5, we can present many equivalent characterizations of the dimension of an individual r.e. real.

The chapter is organized as follows. In Sect. 7.2, we review the previous results on the equivalent characterizations of randomness of an r.e. real. The main result on partial randomness of an r.e. real is presented in Sect. 7.3. We then complete the proof of Theorem 7.5 in Sect. 7.4. In Sect. 7.5 we apply Theorem 7.5 to give many equivalent characterizations of the dimension of an r.e. real. In Sect. 7.6, we investigate further properties of the notion of T-convergence, which plays a crucial role in our characterizations of the partial randomness and dimension of r.e. reals.

7.2   Previous Results on the Randomness of an R.E. Real

In this section, we review the previous results on the randomness of an r.e. real. First we review some important notions on r.e. reals.

Definition 7.1 (*Domination and Ω-Likeness, Solovay* [37], *Chaitin* [10], *Calude et al.* [4]) For any r.e. reals α and β, we say that α *dominates* β if there are computable, increasing sequences $\{a_n\}$ and $\{b_n\}$ of rationals and $c \in \mathbb{N}^+$ such that $\lim_{n\to\infty} a_n = \alpha$, $\lim_{n\to\infty} b_n = \beta$, and $c(\alpha - a_n) \geq \beta - b_n$ for all $n \in \mathbb{N}$. An r.e. real α is called Ω-*like* if it dominates all r.e. reals. $\square$

Solovay [37] showed the following theorem for this notion of domination. For the proof, see also Theorem 4.9 of Calude et al. [4].

Theorem 7.2 (Solovay [37]) *For every r.e. reals α and β, if α dominates β then $K(\beta\restriction_n) \leq K(\alpha\restriction_n) + O(1)$ for all $n \in \mathbb{N}^+$.* $\square$

Thus, the notion of domination for two r.e. reals has an important implication in comparing the program-size complexity of the two reals.

Definition 7.2 (*Universality, Solovay* [37]) A computable, increasing and converging sequence $\{a_n\}$ of rationals is called *universal* if for every computable, increasing and converging sequence $\{b_n\}$ of rationals there exists $c \in \mathbb{N}^+$ such that $c(\alpha - a_n) \geq \beta - b_n$ for all $n \in \mathbb{N}$, where $\alpha := \lim_{n\to\infty} a_n$ and $\beta := \lim_{n\to\infty} b_n$. $\square$

The previous results on the equivalent characterizations of randomness for an r.e. real are summarized in the following theorem.

Theorem 7.3 ([4, 9, 19, 33, 37, 42]) *Let α be an r.e. real with $0 < \alpha < 1$. Then the following conditions are equivalent:*

  (i)   *The real α is weakly Chaitin random.*
  (ii) *The real α is Martin-Löf random.*
  (iii) *The real α is Ω-like.*
  (iv) *For every r.e. real β, it holds that $K(\beta\restriction_n) \leq K(\alpha\restriction_n) + O(1)$ for all $n \in \mathbb{N}^+$.*
  (v)   *There exists an optimal prefix-free machine V such that $\alpha = Z_V(1)$.*
  (vi) *There exists a universal probability m such that $\alpha = \sum_{s\in\{0,1\}^*} m(s)$.*
  (vii) *Every computable, increasing sequence of rationals which converges to α is universal.*
  (viii) *There exists a universal computable, increasing sequence of rationals which converges to α.* $\square$

The historical remark on the proofs of equivalences in Theorem 7.3 is as follows. Schnorr [33] showed that (i) and (ii) are equivalent to each other. Chaitin [9] showed that (v) implies (i). Solovay [37] showed that (v) implies (iii), (iii) implies (iv), and (iii) implies (i). Calude et al. [4] showed that (iii) implies (v), and (v) implies (vii). Kučera and Slaman [19] showed that (ii) implies (vii). Finally, (vi) was inserted in the course of the derivation from (v) to (viii) by Tadaki [42].

7.3 Extension over the Notion of Partial Randomness

In this section, we generalize Theorem 7.3 above over the notion of partial randomness. For that purpose, we first introduce some new notions. Let T denote a real with
$0 < T \le 1$ *throughout the rest of this chapter.* These notions are parameterized by
the real T.

Definition 7.3 (*T-Convergence, Tadaki* [47]) Let $T \in (0, 1]$. An increasing
sequence $\{a_n\}$ of reals is called *T-convergent* if

$$\sum_{n=0}^{\infty} (a_{n+1} - a_n)^T < \infty. \tag{7.1}$$

An r.e. real α is called *T-convergent* if there exists a T-convergent computable,
increasing sequence of rationals which converges to α, i.e., if there exists an increasing sequence $\{a_n\}$ of rationals such that (i) $\{a_n\}$ is T-convergent, (ii) $\{a_n\}$ is computable, and (iii) $\lim_{n\to\infty} a_n = \alpha$. $\qquad\qquad\square$

Note that every increasing and converging sequence of reals is 1-convergent, and
thus every r.e. real is 1-convergent.

In general, based on the following proposition, we can freely switch from
"T-convergent computable, increasing sequence of reals" to "T-convergent computable, increasing sequence of rationals."

Proposition 7.1 *Let $T \in (0, 1]$. For every $\alpha \in \mathbb{R}$, α is a T-convergent r.e. real if and
only if there exists a T-convergent computable, increasing sequence of reals which
converges to α.*

Proof The "only if" part is obvious. We show the "if" part. Suppose that $\{a_n\}$ is
a T-convergent computable, increasing sequence of reals which converges to α.
Then it follows that there exists a computable sequence $\{b_n\}$ of rationals such that
$a_n < b_n < a_{n+1}$ for all $n \in \mathbb{N}$. Obviously, $\{b_n\}$ is an increasing sequence of rationals
which converges to α. Using the inequality $(x + y)^t \le x^t + y^t$ for reals $x, y > 0$ and
$t \in (0, 1]$, we see that $(b_{n+1} - b_n)^T < (a_{n+2} - a_n)^T \le (a_{n+2} - a_{n+1})^T + (a_{n+1} -
a_n)^T$ for each $n \in \mathbb{N}$. Thus, since both $\sum_{n=0}^{\infty}(a_{n+2} - a_{n+1})^T$ and $\sum_{n=0}^{\infty}(a_{n+1} - a_n)^T$
converge, the increasing sequence $\{b_n\}$ of rationals is T-convergent. $\qquad\square$

Based on Proposition 7.1, we can see that the thermodynamic quantities $Z_V(T)$
and $F_V(T)$ for an arbitrary optimal prefix-free machine V are examples of a
T-convergent r.e. real, as the following theorem states.

Theorem 7.4 *Let $T \in (0, 1]$, and let V be an optimal prefix-free machine. If T is
computable, then $Z_V(T)$ and $-F_V(T)$ are T-convergent r.e. reals.*

Proof Let $p_0, p_1, p_2, \ldots$ be any particular recursive enumeration of the r.e. set dom V.
Then $Z_V(T) = \sum_{i=0}^{\infty} 2^{-|p_i|/T}$, and the increasing sequence $\left\{\sum_{i=0}^{n} 2^{-|p_i|/T}\right\}_{n\in\mathbb{N}}$

of reals is T-convergent since $Z_V(1) = \sum_{i=0}^{\infty} 2^{-|p_i|} < 1$. If T is computable, then this sequence of reals is computable. Thus, by Proposition 7.1 we have that $Z_V(T)$ is a T-convergent r.e. real. Based on this, using the relation $F_V(T) = -T \log_2 Z_V(T)$ we have that $-F_V(T)$ is a T-convergent r.e. real. $\square$

Definition 7.4 ($\Omega(T)$-*Likeness, Tadaki* [47]) Let $T \in (0, 1]$. An r.e. real α is called $\Omega(T)$-*like* if it dominates all T-convergent r.e. reals. $\square$

Note that an r.e. real α is $\Omega(1)$-like if and only if α is Ω-like.

Definition 7.5 (T-*Universality, Tadaki* [47]) Let $T \in (0, 1]$. A computable, increasing and converging sequence $\{a_n\}$ of rationals is called T-*universal* if for every T-convergent computable, increasing and converging sequence $\{b_n\}$ of rationals there exists $c \in \mathbb{N}^+$ such that $c(\alpha - a_n) \geq \beta - b_n$ for all $n \in \mathbb{N}$, where $\alpha := \lim_{n \to \infty} a_n$ and $\beta := \lim_{n \to \infty} b_n$. $\square$

Note that a computable, increasing and converging sequence $\{a_n\}$ of rationals is 1-universal if and only if $\{a_n\}$ is universal.

We extend the definition of $Z_V(T)$, which have so far been defined only for an optimal prefix-free machine, over an *arbitrary* prefix-free machine. To be specific, we define $Z_M(T)$ as $\sum_{p \in \mathrm{dom}\, M} 2^{-|p|/T}$ for any prefix-free machine M and real $T > 0$. Note here that $Z_M(T) = 0$ if $\mathrm{dom}\, M = \emptyset$. Obviously, $0 \leq Z_M(T) \leq 1$ holds for every prefix-free machine M and real $T \in (0, 1]$.

Using the notions introduced above, Theorem 7.3 is generalized as follows.

Theorem 7.5 (Characterizations of Partial Randomness for R.E. Real, Tadaki [47]) *Let $T \in (0, 1]$, and let α be an r.e. real with $0 < \alpha < 1$. Suppose that T is computable. Then the following conditions are equivalent:*

  (i)   *The real α is weakly Chaitin T-random.*
 (ii)   *The real α is Martin-Löf T-random.*
(iii)   *The real α is $\Omega(T)$-like.*
 (iv)   *For every prefix-free machine M, the real α dominates $Z_M(T)$.*
  (v)   *There exists an optimal prefix-free machine V such that α dominates $Z_V(T)$.*
 (vi)   *For every T-convergent r.e. real β, it holds that $K(\beta\lceil_n) \leq K(\alpha\lceil_n) + O(1)$ for all $n \in \mathbb{N}^+$.*
(vii)   *For every T-convergent r.e. real $\gamma > 0$, there exist an r.e. real $\beta \geq 0$ and a rational $q > 0$ such that $\alpha = \beta + q\gamma$.*
(viii) *For every optimal prefix-free machine V, there exist an r.e. real $\beta \geq 0$ and a rational $q > 0$ such that $\alpha = \beta + qZ_V(T)$.*
 (ix)   *There exist an optimal prefix-free machine V and an r.e. real $\beta \geq 0$ such that $\alpha = \beta + Z_V(T)$.*
  (x)   *There exists a universal probability m such that $\alpha = \sum_{s \in \{0,1\}^*} m(s)^{1/T}$.*
 (xi)   *Every computable, increasing sequence of rationals which converges to α is T-universal.*
(xii)   *There exists a T-universal computable, increasing sequence of rationals which converges to α.* $\square$

We see that Theorem 7.5 is a massive expansion of Theorem 3.2 in the case where the real α is r.e. with $0 < \alpha < 1$. The condition (ix) of Theorem 7.5 corresponds to the condition (v) of Theorem 7.3. Note, however, that, in the condition (ix) of Theorem 7.5, a non-negative r.e. real β is needed. The reason is as follows: In the case of $\beta = 0$, the possibility that α is weakly Chaitin T'-random with a real $T' > T$ is excluded by the T-compressibility of $Z_V(T)$ imposed by Theorem 5.31 (i). However, this exclusion is not required in the condition (i) of Theorem 7.5.

Theorem 7.5 is proved as follows, partially but substantially based on Theorems 7.6, 7.7, 7.8, and 7.9 which will be proved in the next section.

Proof (of Theorem 7.5) We prove the equivalences in Theorem 7.5 by showing the three paths [A], [B], and [C] of implications below.

[A] The implications (i) $\Rightarrow$ (ii) $\Rightarrow$ (vii) $\Rightarrow$ (viii) $\Rightarrow$ (ix) $\Rightarrow$ (i): First, by Theorem 3.2, (i) implies (ii) obviously. It follows from Theorem 7.6 below that (ii) implies (vii). For the third implication, recall that $Z_V(T)$ is a T-convergent r.e. real for every optimal prefix-free machine V due to Theorem 7.4. Thus, the condition (vii) results in the condition (viii) of Theorem 7.5. Then, it follows from Theorem 7.7 below that (viii) implies (ix). For the fifth implication, let V be an optimal prefix-free machine, and let β be an r.e. real. It is then easy to show that $\beta + Z_V(T)$ dominates $Z_V(T)$ (see the condition 2 of Lemma 4.4 of Calude et al. [4]). It follows from Theorem 7.2 and Theorem 5.31 (i) that the condition (ix) results in the condition (i) of Theorem 7.5.

[B] The implications (vii) $\Rightarrow$ (x) $\Rightarrow$ (xi) $\Rightarrow$ (xii) $\Rightarrow$ (iii) $\Rightarrow$ (vi) $\Rightarrow$ (i): First, it follows from Theorem 7.8 below that (vii) implies (x), and also it follows from Theorem 7.9 below that (x) implies (xi). Obviously, (xi) implies (xii) and (xii) implies (iii). It follows from Theorem 7.2 that (iii) implies (vi). Finally, note that $Z_U(T)$ is a T-convergent r.e. real which is weakly Chaitin T-random by Theorem 7.4 and Theorem 3.1 (i). Thus, by setting β to $Z_U(T)$ in the condition (vi), the condition (vi) results in the condition (i) of Theorem 7.5.

[C] The implications (iii) $\Rightarrow$ (iv) $\Rightarrow$ (v) $\Rightarrow$ (i): First, since T is computable real with $T \in (0, 1]$, we can show that $Z_M(T)$ is a T-convergent r.e. real for every prefix-free machine M, in the same manner as the proof of Theorem 7.4. Thus, the condition (iii) results in the condition (iv) of Theorem 7.5. Obviously, (iv) implies (v). Finally, it follows from Theorem 7.2 and Theorem 5.31 (i) that the condition (v) results in the condition (i) of Theorem 7.5.   □

Theorem 7.5 has many important implications and applications. As we saw in Sect. 3.3, partial randomness and Hausdorff dimension are equivalent notions. The notion of T-convergence appears in the conditions of Theorem 7.5. In the requirement (7.1) which characterizes this notion in Definition 7.3, the real T appears as the exponent of each term in the infinite sum. The form of this type can be seen in the definition of *Hausdorff measure* (see Falconer [15]). This similarity reflects the equivalence between partial randomness and Hausdorff dimension. One of the main applications of Theorem 7.5 is to give many characterizations of the dimension of an individual r.e. real, some of which will be presented in Sect. 7.5.

As another consequence of Theorem 7.5, we can obtain Corollary 7.1 below for example, which follows immediately from the implication (ix) $\Rightarrow$ (vi) of Theorem 7.5 and Theorem 7.4.

Corollary 7.1 *Let $T \in (0, 1]$. Suppose that T is computable. Then, for every two optimal prefix-free machines V and W, it holds that $K(Z_V(T){\restriction}_n) = K(Z_W(T){\restriction}_n) + O(1)$ for all $n \in \mathbb{N}^+$.* $\qquad\square$

Note that the computability of T is necessary for Theorem 7.5 to hold. To see this, let T be a real with $0 < T < 1$ such that $Z_U(T)$ is computable. It follows from Theorem 5.3 that T is not computable (but T is right-computable due to Theorem 5.4). Recall from Theorem 5.2 that the set of such a real T is dense in $(0, 1)$. Let α denote $Z_U(T)$. Then α is an r.e. real with $0 < \alpha < 1$ which satisfies the condition (ix) of Theorem 7.5. However, the real α does not satisfy the condition (i) of Theorem 7.5, since α is computable and therefore not weakly Chaitin T-random.

The notion of T-convergence has many interesting properties, in addition to the properties which we saw above. In Sect. 7.6, we investigate further properties of the notion.

7.4 The Completion of the Proof of Theorem 7.5

In this section, we prove several theorems needed to complete the proof of Theorem 7.5. First, we prove Lemma 7.1 below. Lemma 7.1 and Theorem 7.10 below can be proved, through a generalization of the techniques used in the proof of Theorem 2.1 of Kučera and Slaman [19] over the notion of partial randomness.

Lemma 7.1 *Let $T \in (0, 1]$. Let α be an r.e. real, and let $\{d_n\}$ be a computable sequence of positive rationals such that $\sum_{n=0}^{\infty} d_n{}^T \le 1$. If α is Martin-Löf T-random, then for every $\varepsilon > 0$ there exist a computable, increasing sequence $\{a_n\}$ of rationals and a rational $q > 0$ such that $a_{n+1} - a_n > qd_n$ for every $n \in \mathbb{N}$, $a_0 > \alpha - \varepsilon$, and $\alpha = \lim_{n \to \infty} a_n$.*

Proof We first choose any particular $c \in \mathbb{N}^+$ with $cT \ge 1$, and then define a function $f \colon \mathbb{N} \times \mathbb{N}^+ \to \mathbb{N}^+$ by $f(n, i) = \lceil -\log_2 d_n + c(i + 1) \rceil$. Since $\{d_n\}$ is a computable sequence of rationals, it is easy to see that $f \colon \mathbb{N} \times \mathbb{N}^+ \to \mathbb{N}^+$ is a total recursive function. Let ε be an arbitrary positive real. Then, since α is an r.e. real, there exists a computable, increasing sequences $\{b_n\}$ of rationals such that $b_0 > \alpha - \varepsilon$ and $\alpha = \lim_{n \to \infty} b_n$. We construct a Martin-Löf T-test $\mathscr{C}$ by enumerating $\mathscr{C}_i$ for each $i \in \mathbb{N}^+$, where $\mathscr{C}_i$ denotes the set $\{ s \mid (i, s) \in \mathscr{C} \}$. During the enumeration of $\mathscr{C}_i$ we simultaneously construct a sequence $\{a(i)_n\}_n$ of rationals.

For that purpose, we introduce some notation. For each $s \in \{0, 1\}^*$ with $s \neq \lambda$, we define a finite subset $G(s)$ of $\{0, 1\}^*$ in the following manner: If the finite binary string s consists of only ones, we set $G(s) := \{s\}$. But if not (i.e., if the string s contains a zero), we set $G(s) := \{s, s + 1\}$ where $s + 1$ denotes the finite binary string of the same length as s, obtained by adding 1 to s under the common identification of a finite binary string with a dyadic integer. For example, $G(111) = \{111\}$ and $G(0011) = \{0011, 0100\}$. Moreover, for each subset S of $\{0, 1\}^*$ we denote by $[S]^{\prec}$ the set of all reals γ such that $\gamma\lceil_k \in S$ for some $k \in \mathbb{N}$.

Now, for each $i \in \mathbb{N}^+$, the enumeration of $\mathscr{C}_i$ is performed as follows: Initially, we set $\mathscr{C}_i := \emptyset$ and then specify $a(i)_0$ by $a(i)_0 := b_0$. In general, whenever $a(i)_n$ is specified as $a(i)_n := b_m$, we update $\mathscr{C}_i$ by

$$\mathscr{C}_i := \mathscr{C}_i \cup G(a(i)_n\lceil_{f(n,i)}),$$

and calculate $b_{m+1}, b_{m+2}, b_{m+3}, \cdots$ one by one in this order. During the calculation, if we find m_1 such that $m_1 > m$ and $b_{m_1} \notin \left[G(a(i)_n\lceil_{f(n,i)})\right]^{\prec}$, then we specify $a(i)_{n+1}$ by $a(i)_{n+1} := b_{m_1}$ and we repeat this procedure for $n + 1$.

For the completed $\mathscr{C}$ through the above procedure, we see for each $i \in \mathbb{N}^+$ that

$$\sum_{s \in \mathscr{C}_i} 2^{-T|s|} = \sum_n \sum_{s \in G(a(i)_n\lceil_{f(n,i)})} 2^{-T|s|} \le \sum_n 2 \cdot 2^{-Tf(n,i)} \le 2 \sum_n d_n^T 2^{-cT(i+1)}$$

$$\le 2 \sum_n d_n^T 2^{-(i+1)} = 2^{-i} \sum_n d_n^T \le 2^{-i}.$$

Here, the five sums on n may be finite or infinite, depending on whether the set $\mathscr{C}_i$ is finite or infinite. Thus, since $\mathscr{C}$ is an r.e. set, $\mathscr{C}$ is a Martin-Löf T-test. Therefore, since α is Martin-Löf T-random, there exists $k \in \mathbb{N}^+$ such that, for every $\ell \in \mathbb{N}^+$, it holds that $\alpha\lceil_\ell \notin \mathscr{C}_k$. Recall here that $\alpha = \lim_{n\to\infty} b_n$ and the sequence $\{b_n\}$ is increasing. Thus, in the above procedure for enumerating $\mathscr{C}_k$, for each $n \in \mathbb{N}$ we ever find m_1 such that $m_1 > m$ and $b_{m_1} \notin \left[G(a(k)_n\lceil_{f(n,k)})\right]^{\prec}$. Actually, this b_{m_1} is greater than any real in $\left[G(a(k)_n\lceil_{f(n,k)})\right]^{\prec}$ and therefore $b_{m_1} > a(k)_n + 2^{-f(n,k)}$.

Hence, $\mathscr{C}_k$ is constructed as an infinite set and also $\{a(k)_n\}_n$ is constructed as an infinite sequence of rationals. Furthermore, $a(k)_{n+1} > a(k)_n + 2^{-f(n,k)}$ holds for all $n \in \mathbb{N}$. Therefore, for each $n \in \mathbb{N}$, we have that $a(k)_{n+1} - a(k)_n > 2^{-f(n,k)} > 2^{-c(k+1)-1}d_n$. Since $\{a(k)_n\}_n$ is a subsequence of $\{b_n\}$, it follows that the sequence $\{a(k)_n\}_n$ is increasing and $\alpha = \lim_{n\to\infty} a(k)_n$. In addition, we have $a(k)_0 > \alpha - \varepsilon$ since $a(k)_0 = b_0$. This completes the proof. $\square$

Theorem 7.6 *Let $T \in (0, 1]$. For every r.e. real $\alpha > 0$, if α is Martin-Löf T-random, then for every T-convergent r.e. real $\gamma > 0$ there exist an r.e. real $\beta > 0$ and a rational $q > 0$ such that $\alpha = \beta + q\gamma$.*

Proof Suppose that γ is an arbitrary T-convergent r.e. real with $\gamma > 0$. Then there exists a T-convergent computable, increasing sequence $\{c_n\}$ of rationals which converges to γ. Since $\gamma > 0$, without loss of generality we can assume that $c_0 = 0$. We choose any particular rational $\varepsilon > 0$ such that

$$\sum_{n=0}^{\infty} (c_{n+1} - c_n)^T \le \left(\frac{1}{\varepsilon}\right)^T.$$

Such ε exists since the sequence $\{c_n\}$ is T-convergent. It follows that

$$\sum_{n=0}^{\infty} \left[\varepsilon(c_{n+1} - c_n)\right]^T \leq 1.$$

Note that the sequence $\{\varepsilon(c_{n+1} - c_n)\}_n$ is a computable sequence of positive rationals. Thus, since α is r.e. and Martin-Löf T-random by the assumption, it follows from Lemma 7.1 that there exist a computable, increasing sequence $\{a_n\}$ of rationals and a rational $r > 0$ such that $a_{n+1} - a_n > r\varepsilon(c_{n+1} - c_n)$ for every $n \in \mathbb{N}$, $a_0 > 0$, and $\alpha = \lim_{n\to\infty} a_n$. We then define a sequence $\{b_n\}$ of positive rationals by $b_n = a_{n+1} - a_n - r\varepsilon(c_{n+1} - c_n)$. It follows that $\{b_n\}$ is a computable sequence of rationals and $\sum_{n=0}^{\infty} b_n$ converges to $\alpha - a_0 - r\varepsilon(\gamma - c_0)$. Thus we have $\alpha = a_0 + \sum_{n=0}^{\infty} b_n + r\varepsilon\gamma$, where $a_0 + \sum_{n=0}^{\infty} b_n$ is a positive r.e. real. This completes the proof. $\square$

Theorem 7.7 *Let $T \in (0, 1]$. Suppose that T is computable. For every real α, if for every optimal prefix-free machine V there exist an r.e. real $\beta \geq 0$ and a rational $q > 0$ such that $\alpha = \beta + qZ_V(T)$, then there exist an optimal prefix-free machine W and an r.e. real $\gamma \geq 0$ such that $\alpha = \gamma + Z_W(T)$.*

Proof First, we choose any particular optimal prefix-free machine V. Then, by the assumption, there exist an r.e. real $\beta \geq 0$ and a rational $q > 0$ such that $\alpha = \beta + qZ_V(T)$. We choose any particular $n \in \mathbb{N}$ with $q > 2^{-n/T}$. We then define a partial function $W: \{0, 1\}^* \to \{0, 1\}^*$ by the conditions that (i) $\text{dom } W = \{0^n p \mid p \in \text{dom } V\}$ and (ii) $W(0^n p) = V(p)$ for every $p \in \text{dom } V$. Since $\text{dom } W$ is a prefix-free set, it follows that W is a prefix-free machine. It is then easy to see that $K_W(s) = K_V(s) + n$ for every $s \in \{0, 1\}^*$. Therefore, since V is an optimal prefix-free machine, W is also an optimal prefix-free machine. It follows that $Z_W(T) = 2^{-n/T}Z_V(T)$. Thus we have $\alpha = \beta + (q - 2^{-n/T})Z_V(T) + Z_W(T)$. On the other hand, since T is computable, $\beta + (q - 2^{-n/T})Z_V(T)$ is an r.e. real. This completes the proof. $\square$

Theorem 7.8 *Let $T \in (0, 1]$. Suppose that T is computable. For every real $\alpha \in (0, 1)$, if for every T-convergent r.e. real $\gamma > 0$ there exist an r.e. real $\beta \geq 0$ and a rational $q > 0$ such that $\alpha = \beta + q\gamma$, then there exists a universal probability m such that $\alpha = \sum_{s \in \{0,1\}^*} m(s)^{1/T}$.*

Proof First, consider a set $R := \{(n, s) \in \mathbb{N} \times \{0, 1\}^* \mid n > K(s)\}$. Since R is an infinite r.e. set, there exists an injective total recursive function $f: \mathbb{N} \to \mathbb{N} \times \{0, 1\}^*$ such that $f(\mathbb{N}) = R$. Let $f_1: \mathbb{N} \to \mathbb{N}$ and $f_2: \mathbb{N} \to \{0, 1\}^*$ be total recursive functions such that $f(k) = (f_1(k), f_2(k))$ for all $k \in \mathbb{N}$. Then

$$\sum_{k=0}^{\infty} 2^{-f_1(k)} = \sum_{n > K(s)} 2^{-n} = \sum_{s \in \{0,1\}^*} \sum_{n=K(s)+1}^{\infty} 2^{-n} = \sum_{s \in \{0,1\}^*} 2^{-K(s)} \leq \Omega < 1,$$

where the second sum is over all $(n, s) \in \mathbb{N} \times \{0, 1\}^*$ with $n > K(s)$. Thus, it follows from Theorem 2.3 that there exists a prefix-free machine V which satisfies the following two conditions:

(i) For every $n \in \mathbb{N}$ and $s \in \{0, 1\}^*$, it holds that

$$\#\{ p \in \{0, 1\}^* \mid |p| = n \ \& \ V(p) = s \} = \#\{ k \in \mathbb{N} \mid f_1(k) = n \ \& \ f_2(k) = s \}.$$

(ii) $K_V(s) = \min\{ f_1(k) \mid f_2(k) = s \}$ for every $s \in \{0, 1\}^*$.

It is then easy to see that (i) $K_V(s) = K(s) + 1$ for every $s \in \{0, 1\}^*$ and (ii) for every $n \in \mathbb{N}$ and $s \in \{0, 1\}^*$, if $n > K(s)$ then there exists a unique $p \in \{0, 1\}^*$ such that $|p| = n$ and $V(p) = s$. Thus, the prefix-free machine V is optimal and

$$Z_V(T) = \sum_{s \in \{0,1\}^*} \sum_{n=K(s)+1}^{\infty} 2^{-n/T} = \frac{1}{2^{1/T} - 1} \sum_{s \in \{0,1\}^*} 2^{-K(s)/T}. \tag{7.2}$$

By Theorem 7.4, note that $Z_V(T)$ is a T-convergent r.e. real. Hence, by the assumption, there exist an r.e. real $\beta \geq 0$ and a rational $q > 0$ such that $\alpha = \beta + q Z_V(T)$. We choose any particular rational $\varepsilon > 0$ such that $\varepsilon \leq 1 - \alpha^T$ and $\varepsilon^{1/T} < q/(2^{1/T} - 1)$. It follows from (7.2) that

$$\alpha = \beta + \frac{q}{2^{1/T} - 1} 2^{-K(\lambda)/T} + \left(\frac{q}{2^{1/T} - 1} - \varepsilon^{1/T} \right) \sum_{s \neq \lambda} 2^{-K(s)/T}$$
$$+ \sum_{s \neq \lambda} \left(\varepsilon 2^{-K(s)} \right)^{1/T}. \tag{7.3}$$

Let δ be the sum of the first, second, and third terms on the right-hand side of (7.3). Then, since T is computable, δ is an r.e. real. We define a function $m \colon \{0, 1\}^* \to (0, \infty)$ by $m(s) = \delta^T$ if $s = \lambda$; $m(s) = \varepsilon 2^{-K(s)}$ otherwise. Since $\sum_{s \neq \lambda} \varepsilon 2^{-K(s)} < \varepsilon$ and $\delta^T < \alpha^T \leq 1 - \varepsilon$, we see that $\sum_{s \in \{0,1\}^*} m(s) < \delta^T + \varepsilon < 1$. Since T is right-computable, δ^T is an r.e. real. Therefore, since $2^{-K(s)}$ is a lower-computable semi-measure by Theorem 2.1, m is also a lower-computable semi-measure. Thus, since $2^{-K(s)}$ is a universal probability by Theorem 2.1 again and $\delta^T > 0$, we have that m is a universal probability. On the other hand, it follows from (7.3) that $\alpha = \sum_{s \in \{0,1\}^*} m(s)^{1/T}$. This completes the proof. $\qquad \square$

Theorem 7.9 below is obtained by generalizing the proofs of Solovay [37] and Theorem 6.4 of Calude et al. [4].

Theorem 7.9 *Let $T \in (0, 1]$. Suppose that T is computable. For every $\alpha \in (0, 1)$, if there exists a universal probability m such that $\alpha = \sum_{s \in \{0,1\}^*} m(s)^{1/T}$, then every computable, increasing sequence of rationals which converges to α is T-universal.*

Proof Suppose that $\{a_n\}$ is an arbitrary computable, increasing sequence of rationals which converges to $\alpha = \sum_{s \in \{0,1\}^*} m(s)^{1/T}$. Since m is a lower-computable semi-measure and T is left-computable, there exists a total recursive function $f : \mathbb{N} \to \mathbb{N}^+$ such that, for every $n \in \mathbb{N}$, it holds that $f(n) < f(n+1)$ and

$$\sum_{k=0}^{f(n)-1} m(k)^{1/T} \geq a_n. \tag{7.4}$$

Recall here that we identify $\{0, 1\}^*$ with $\mathbb{N}$. We then define a total recursive function $g : \mathbb{N} \to \mathbb{N}$ by $g(k) = \min\{n \in \mathbb{N} \mid k \leq f(n)\}$. It follows that $g(f(n)) = n$ for every $n \in \mathbb{N}$ and $\lim_{k \to \infty} g(k) = \infty$.

Let $\{b_n\}$ be an arbitrary T-convergent computable, increasing and converging sequence of rationals. We choose any particular $d \in \mathbb{N}^+$ with $\sum_{n=0}^{\infty}(b_{n+1} - b_n)^T \leq d$, and then define a function $r : \mathbb{N} \to [0, \infty)$ by $r(k) = (b_{g(k+1)} - b_{g(k)})^T/d$. Since (i) T is computable and (ii) $g(k+1)$ equals to $g(k)$ or $g(k) + 1$ for every $k \in \mathbb{N}$, it follows that r is a lower-computable semi-measure. Thus, since m is a universal probability, there exists $c \in \mathbb{N}^+$ such that, for every $k \in \mathbb{N}$, it holds that $cm(k) \geq r(k)$. It follows from (7.4) that, for each $n \in \mathbb{N}$,

$$(cd)^{1/T}(\alpha - a_n) \geq d^{1/T} \sum_{k=f(n)}^{\infty} (cm(k))^{1/T} \geq \sum_{k=f(n)}^{\infty} (b_{g(k+1)} - b_{g(k)}) = \beta - b_n,$$

where $\beta = \lim_{n \to \infty} b_n$. This completes the proof. $\qquad\square$

Note that, using Lemma 7.1, we can directly show that the condition (ii) implies the condition (iii) in Theorem 7.5 without assuming the computability of T, as follows.

Theorem 7.10 *Let $T \in (0, 1]$. For every r.e. real α, if α is Martin-Löf T-random, then α is $\Omega(T)$-like.*

Proof Suppose that β is an arbitrary T-convergent r.e. real. Then there is a T-convergent computable, increasing sequence $\{b_n\}$ of rationals which converges to β. Since $\{b_n\}$ is T-convergent, without loss of generality we can assume that $\sum_{n=0}^{\infty}(b_{n+1} - b_n)^T \leq 1$. Since α is r.e. and Martin-Löf T-random by the assumption, it follows from Lemma 7.1 that there exist a computable, increasing sequence $\{a_n\}$ of rationals and a rational $q > 0$ such that $a_{n+1} - a_n > q(b_{n+1} - b_n)$ for every $n \in \mathbb{N}$ and $\alpha = \lim_{n \to \infty} a_n$. It is then easy to see that $\alpha - a_n > q(\beta - b_n)$ for every $n \in \mathbb{N}$. Therefore α dominates β. This completes the proof. $\qquad\square$

7.5 Characterizations of the Dimension of an R.E. Real

In this section we apply Theorem 7.5 to give many characterizations of dimension for an individual r.e. real. As referred to in Sect. 3.3, through the works [40, 41], we introduced the notions of six "algorithmic dimensions", 1st, 2nd, 3rd, 4th, upper, and lower algorithmic dimensions as fractal dimensions for a subset F of N-dimensional Euclidean space $\mathbb{R}^N$. These notions are defined based on the notion of partial randomness and compression rate by means of program-size complexity.

Among the six algorithmic dimensions, in particular, the notion of *lower algorithmic dimension* for a subset of $\mathbb{R}^N$ is given in Definition 3.7. The lower algorithmic dimension exists for an arbitrary subset of $\mathbb{R}^N$ with an arbitrary $N \in \mathbb{N}^+$, as we can see from Definition 3.7. Thus, of course, we can consider the lower algorithmic dimension for an individual real. Specifically, for each $\alpha \in \mathbb{R}$ we can consider the lower algorithmic dimension $\underline{\dim}_A\{\alpha\}$ of the singleton $\{\alpha\}$, which is given by

$$\underline{\dim}_A\{\alpha\} = \liminf_{n \to \infty} \frac{K(\alpha\restriction_n)}{n}. \tag{7.5}$$

For example, $\underline{\dim}_A\{Z(T)\} = T$ holds for every computable real $T \in (0, 1]$, due to Theorem 3.1 (i).

Independently of us, Lutz [23] introduced the notion of constructive dimension of an individual real α using the notion of lower semicomputable s-supergale with $s \in [0, \infty)$, and then Mayordomo [25] showed that, for every real α, the constructive dimension of α equals to the right-hand side of (7.5). Thus, the constructive dimension of α is precisely the lower algorithmic dimension $\underline{\dim}_A\{\alpha\}$ of α for every real α.

Using Lemma 7.2 below, we can convert each of all the conditions in Theorem 7.5 into a characterization of the lower algorithmic dimension $\underline{\dim}_A\{\alpha\}$ for any r.e. real α.

Lemma 7.2 (Tadaki [41, 47]) *Let $\alpha \in \mathbb{R}$. For every $t \in [0, 1]$, α is weakly Chaitin t-random if $t < \underline{\dim}_A\{\alpha\}$, and α is not weakly Chaitin t-random if $t > \underline{\dim}_A\{\alpha\}$.*

Proof Let $t \in [0, 1]$. First, assume that $t < \underline{\dim}_A\{\alpha\}$. Then, $t < K(\alpha\restriction_n)/n$ for all sufficiently large $n \in \mathbb{N}^+$, and therefore we have that α is weakly Chaitin t-random. On the other hand, assume that α is weakly Chaitin t-random. Then we see that $t \le \liminf_{n \to \infty} K(\alpha\restriction_n)/n = \underline{\dim}_A\{\alpha\}$. Thus, if $t > \underline{\dim}_A\{\alpha\}$ then α is not weakly Chaitin t-random. This completes the proof. $\square$

Thus, in Theorem 7.11 below, each characterization of dimension is obtained by applying Lemma 7.2 to the corresponding condition in Theorem 7.5. We here denote by $\mathbb{R}_c$ the set of all computable reals, and we interpret the supremum of the empty set as 0.

Theorem 7.11 (Characterizations of Dimension for R.E. Real, Tadaki [47]) *Let α be an r.e. real. Then the following hold.*

(i) $\underline{\dim}_A\{\alpha\}$ *equals to the supremum of $T \in (0, 1]$ such that α is weakly Chaitin T-random.*

(ii) $\underline{\dim}_A\{\alpha\}$ *equals to the supremum of $T \in (0, 1]$ such that α is Martin-Löf T-random.*

(iii) $\underline{\dim}_A\{\alpha\}$ *equals to the supremum of* $T \in (0, 1]$ *such that α is $\Omega(T)$-like.*

(iv) $\underline{\dim}_A\{\alpha\}$ *equals to the supremum of* $T \in (0, 1] \cap \mathbb{R}_c$ *such that α dominates* $Z_M(T)$ *for every prefix-free machine M.*

 (v) $\underline{\dim}_A\{\alpha\}$ *equals to the supremum of* $T \in (0, 1] \cap \mathbb{R}_c$ *such that α dominates* $Z_V(T)$ *for some optimal prefix-free machine V.*

(vi) $\underline{\dim}_A\{\alpha\}$ *equals to the supremum of* $T \in (0, 1]$ *such that* $K(\beta\restriction_n) \leq$ $K(\alpha\restriction_n) + O(1)$ *for every T-convergent r.e. real β.*

(vii) $\underline{\dim}_A\{\alpha\}$ *equals to the supremum of* $T \in (0, 1]$ *such that, for every* T-*convergent r.e. real $\gamma > 0$, there exist an r.e. real $\beta \geq 0$ and a rational* $q > 0$ *for which $\alpha = \beta + q\gamma$.*

(viii) $\underline{\dim}_A\{\alpha\}$ *equals to the supremum of* $T \in (0, 1] \cap \mathbb{R}_c$ *such that, for every optimal prefix-free machine V, there exist an r.e. real $\beta \geq 0$ and a rational $q > 0$ for which $\alpha = \beta + qZ_V(T)$.*

(ix) $\underline{\dim}_A\{\alpha\}$ *equals to the supremum of* $T \in (0, 1] \cap \mathbb{R}_c$ *such that there exist an optimal prefix-free machine V and an r.e. real $\beta \geq 0$ for which $\alpha =$* $\beta + Z_V(T)$.

 (x) *If $0 < \alpha < 1$, then $\underline{\dim}_A\{\alpha\}$ equals to the supremum of $T \in (0, 1] \cap \mathbb{R}_c$ such that $\alpha = \sum_{s \in \{0,1\}^*} m(s)^{1/T}$ for some universal probability m.*

(xi) $\underline{\dim}_A\{\alpha\}$ *equals to the supremum of* $T \in (0, 1]$ *such that every computable, increasing sequence of rationals which converges to α is T-universal.*

(xii) $\underline{\dim}_A\{\alpha\}$ *equals to the supremum of* $T \in (0, 1]$ *such that there exists a* T-*universal computable, increasing sequence of rationals which converges to α.* $\qquad\square$

Recall that Theorem 7.5 holds for a *computable* real T with $0 < T \leq 1$. However, the computability of T is eliminated in the characterizations (i), (ii), (iii), (vi), (vii), (xi), and (xii) of dimension in Theorem 7.11. This is possible due to Proposition 7.2 below and the fact that $\mathbb{R}_c$ is dense in $\mathbb{R}$.[2]

Proposition 7.2 *Let T_1 and T_2 be arbitrary reals with $0 < T_1 \leq T_2 \leq 1$. Then the following hold:*

  (i) *If a real is Martin-Löf T_2-random then it is Martin-Löf T_1-random.*
 (ii) *If an increasing sequence of reals is T_1-convergent then it is T_2-convergent.*
(iii) *If an r.e. real is T_1-convergent then it is T_2-convergent.*
(iv) *If an r.e. real is $\Omega(T_2)$-like then it is $\Omega(T_1)$-like.*
 (v) *If a computable, increasing and converging sequence of rationals is* T_2-*universal then it is T_1-universal.* $\qquad\square$

---

[2]For the characterization (i) of Theorem 7.11, the computability of T is not required simply due to the form of Lemma 7.2 itself.

7.6 Further Properties of T-Convergence

In this section, we investigate further properties of the notion of T-convergence. First, as one of the applications of Theorem 7.5, we have the following result.

Theorem 7.12 *Let* $T \in (0, 1]$. *Suppose that* T *is computable. For every r.e. real* α, *if* α *is* T-*convergent, then* α *is* T-*compressible.*

Proof Using (ix) $\Rightarrow$ (vi) of Theorem 7.5 we see that, for every T-convergent r.e. real α, it holds that $K(\alpha\!\upharpoonright_n) \le K(Z_U(T)\!\upharpoonright_n) + O(1)$ for all $n \in \mathbb{N}^+$. It follows from Theorem 3.1 (i) that α is T-compressible for every T-convergent r.e. real α. $\square$

In the case of $0 < T < 1$, the T-compressibility in Theorem 7.12 can be strengthened to the *strict* T-*compressibility*, based also on Theorem 7.5, as we will see in Theorem 8.14 in Sect. 8.5. Also in this case, the converse of Theorem 7.12 does not hold. This fact can be shown by Theorem 7.13 below, together with the T-compressibility of $W(Q, T)$ due to Theorem 3.4 (i).

Theorem 7.13 *Let* $T \in (0, 1)$. *Suppose that* T *is computable. Then, for every computable real* $Q > 0$, *the r.e. real* $W(Q, T)$ *is not* T-*convergent.* $\square$

In Sect. 8.5, in a general framework, we will show that any Chaitin T-random r.e. real cannot be T-convergent if T is computable with $0 < T < 1$, as a corollary of Theorem 8.14 mentioned above. Based on this, we can easily show that neither $E(T)$, $S(T)$, $C(T)$ nor $W(Q, T)$ (with a computable positive real Q) can be T-convergent if T is computable with $0 < T < 1$ (see Theorem 8.16). In Theorem 7.13, however, we give a direct proof of this fact especially for $W(Q, T)$, making immediate use of Theorem 7.5.

Now, in order to prove Theorem 7.13, the following lemma is useful.

Lemma 7.3 *Let* $T \in (0, 1]$.

(i) *If* $\{a_n\}$ *is a* T-*convergent increasing sequence of reals, then every subsequence of the sequence* $\{a_n\}$ *is also* T-*convergent.*

(ii) *Let* α *be a* T-*convergent r.e. real. If* $\{a_n\}$ *is a computable, increasing sequence of rationals converging to* α, *then there exists a subsequence* $\{a'_n\}$ *of the sequence* $\{a_n\}$ *such that* $\{a'_n\}$ *is a* T-*convergent computable, increasing sequence of rationals converging to* α.

Proof (i) Let $f \colon \mathbb{N} \to \mathbb{N}$ such that $f(n) < f(n+1)$ for all $n \in \mathbb{N}$. Then, using repeatedly the inequality $(x + y)^t \le x^t + y^t$ for reals $x, y > 0$ and $t \in (0, 1]$, we have

$$\left(a_{f(n+1)} - a_{f(n)}\right)^T = \left[\sum_{k=f(n)}^{f(n+1)-1} (a_{k+1} - a_k)\right]^T \le \sum_{k=f(n)}^{f(n+1)-1} (a_{k+1} - a_k)^T$$

for each $n \in \mathbb{N}$. It follows that

$$\sum_{n=0}^{m} \left(a_{f(n+1)} - a_{f(n)}\right)^T \leq \sum_{k=f(0)}^{f(m+1)-1} \left(a_{k+1} - a_k\right)^T$$

for all $m \in \mathbb{N}$. Since $\{a_n\}$ is T-convergent, we have that the subsequence $\{a_{f(n)}\}$ of $\{a_n\}$ is also T-convergent.

(ii) We choose any particular T-convergent computable, increasing sequence $\{b_n\}$ of rationals converging to α. It is then easy to show that there exist total recursive functions $g : \mathbb{N} \to \mathbb{N}$ and $h : \mathbb{N} \to \mathbb{N}$ such that, for every $n \in \mathbb{N}$, it holds that (i) $g(n) < g(n+1)$, (ii) $h(n) < h(n+1)$, and (iii) $b_{g(n)} < a_{h(n)} < b_{g(n+1)}$. Using the inequality $(x + y)^t \leq x^t + y^t$ for reals $x, y > 0$ and $t \in (0, 1]$, we then see that

$$\left(a_{h(n+1)} - a_{h(n)}\right)^T < \left(b_{g(n+2)} - b_{g(n)}\right)^T \leq \left(b_{g(n+2)} - b_{g(n+1)}\right)^T + \left(b_{g(n+1)} - b_{g(n)}\right)^T$$

for each $n \in \mathbb{N}$. It follows from Lemma 7.3 (i) that the subsequence $\{b_{g(n)}\}$ of $\{b_n\}$ is T-convergent. Thus, we have that the subsequence $\{a_{h(n)}\}$ of $\{a_n\}$ is a T-convergent computable, increasing sequence of rationals converging to α. $\qquad\square$

Then, the proof Theorem 7.13 is given as follows.

Proof (of Theorem 7.13) Let Q be an arbitrary computable positive real. First, since T is computable with $0 < T < 1$, we recall from Theorem 3.4 (i) that $W(Q, T)$ is an r.e. real. We choose any particular recursive enumeration $p_0, p_1, p_2, \ldots$ of the r.e. set dom U. Then we have $W(Q, T) = \sum_{i=0}^{\infty} |p_i|^Q 2^{-|p_i|/T}$. Since T and Q are computable, it is easy to show that there exists a computable, increasing sequence $\{a_n\}$ of rationals such that

$$\sum_{i=0}^{n-1} |p_i|^Q 2^{-|p_i|/T} < a_n < \sum_{i=0}^{n} |p_i|^Q 2^{-|p_i|/T} \tag{7.6}$$

for all $n \in \mathbb{N}^+$. Obviously, $\{a_n\}$ is an increasing sequence of rationals converging to $W(Q, T)$.

Now, let us assume contrarily that $W(Q, T)$ is T-convergent. Then it follows from Lemma 7.3 (ii) that there exists a total recursive function $f : \mathbb{N} \to \mathbb{N}$ such that $f(n) < f(n+1)$ for all $n \in \mathbb{N}$, and $\{a_{f(n)}\}$ is a T-convergent computable, increasing sequence of rationals converging to $W(Q, T)$. On the other hand, since T is computable, it is easy to show that there exists a computable, increasing sequence $\{b_n\}$ of rationals such that

$$\sum_{i=0}^{f(n)} 2^{-|p_i|/T} < b_n < \sum_{i=0}^{f(n+1)} 2^{-|p_i|/T} \tag{7.7}$$

for all $n \in \mathbb{N}$. Obviously, $\{b_n\}$ converges to $Z(T)$. Since U is an optimal prefix-free machine, using (ix) $\Rightarrow$ (xi) of Theorem 7.5, we see that there exists $c \in \mathbb{N}^+$ such that $c(Z(T) - b_n) \geq W(Q, T) - a_{f(n)}$ for all $n \in \mathbb{N}$. It follows from (7.6) and (7.7) that

$$c \sum_{i=f(n)+1}^{\infty} 2^{-|p_i|/T} > \sum_{i=f(n)+1}^{\infty} |p_i|^Q 2^{-|p_i|/T}$$

for all $n \in \mathbb{N}^+$. Therefore, we have that

$$\sum_{i=f(n)+1}^{\infty} (c - |p_i|^Q) 2^{-|p_i|/T} > 0 \qquad (7.8)$$

for all $n \in \mathbb{N}^+$. On the other hand, it is easy to see that $\lim_{i \to \infty} |p_i|^Q = \infty$. Therefore, since $\lim_{n \to \infty} f(n) = \infty$, there exists $n_0 \in \mathbb{N}^+$ such that, for all $i \in \mathbb{N}$, if $i \geq f(n_0) + 1$ then $|p_i|^Q \geq c$. Thus, by setting n to n_0 in (7.8), we have a contradiction. This completes the proof. $\qquad \square$

Let T_1 and T_2 be arbitrary computable reals with $0 < T_1 < T_2 < 1$, and let V be an arbitrary optimal prefix-free machine. By Theorem 3.1 (i) and Theorem 7.12, we see that the r.e. real $Z_V(T_2)$ is not T_1-convergent and therefore every computable, increasing sequence of rationals which converges to $Z_V(T_2)$ is not T_1-convergent. At this point, conversely, the following question arises naturally: Is there any computable, increasing sequence of rationals which converges to $Z_V(T_1)$ and which is not T_2-convergent? We can answer this question affirmatively in the form of Theorem 7.14 below, if we can choose the optimal prefix-free machine V arbitrarily depending on T_1 and T_2.

Theorem 7.14 *Let T_1 and T_2 be arbitrary computable reals with $0 < T_1 < T_2 < 1$. Then there exist an optimal prefix-free machine V and a computable, increasing sequence $\{a_n\}$ of rationals such that (i) $Z_V(T_1) = \lim_{n \to \infty} a_n$, (ii) $\{a_n\}$ is T-convergent for every $T \in (T_2, 1]$, and (iii) $\{a_n\}$ is not T-convergent for every $T \in (0, T_2]$.*

Proof First, we choose any particular computable, increasing sequence $\{c_n\}$ of reals such that (i) $\{c_n\}$ converges to a computable real $\gamma > 0$, (ii) $\{c_n\}$ is T-convergent for every $T \in (T_2, 1]$, and (iii) $\{c_n\}$ is not T-convergent for every $T \in (0, T_2]$. Such a sequence $\{c_n\}$ can be obtained, for example, in the following manner.

Let $\{c_n\}$ be an increasing sequence of reals with

$$c_n = \sum_{k=1}^{n+1} \left(\frac{1}{k} \right)^{1/T_2}.$$

Since T_2 is a computable real with $0 < T_2 < 1$, it is easy to see that $\{c_n\}$ is a computable sequence of reals which converges to a computable positive real. We can also easily see that $\{c_n\}$ is T-convergent for every $T \in (T_2, 1]$, and $\{c_n\}$ is not T-convergent for every $T \in (0, T_2]$. Thus, this sequence $\{c_n\}$ has the properties (i), (ii), and (iii) desired above.

We choose any particular rational r with $0 < r < 1/\gamma$, and denote $r\gamma$ by β. Obviously, β is a computable real with $0 < \beta < 1$. We denote $2^{1/T_1}$ by b. Then

$1 < b$. We can then effectively expand β to the base-b; that is, Property 7.1 below holds for the pair of β and b.

Property 7.1 *There exists a total recursive function* $f: \mathbb{N}^+ \to \mathbb{N}$ *such that* $f(k) \leq \lceil b \rceil - 1$ *for all* $k \in \mathbb{N}^+$ *and* $\beta = \sum_{k=1}^{\infty} f(k)b^{-k}$.

This can be possible since both β and b are computable. The detail is as follows. In the case where Property 7.2 below holds for the pair of β and b, Property 7.1 holds, obviously.

Property 7.2 *There exist* $m \in \mathbb{N}^+$ *and a function* $g: \{1, 2, \ldots, m\} \to \mathbb{N}$ *such that* $g(k) \leq \lceil b \rceil - 1$ *for all* $k \in \{1, 2, \ldots, m\}$ *and* $\beta = \sum_{k=1}^{m} g(k)b^{-k}$.

Thus, in what follows, we assume that Property 7.2 does not hold. In this case, we construct the total recursive function $f: \mathbb{N}^+ \to \mathbb{N}$ by calculating

$$f(1), f(2), f(3), \ldots, f(m), \ldots$$

one by one in this order, based on recursion on stages m. We start with stage 1 and follow the instructions below. Note that the sum $\sum_{k=1}^{m-1} f(k)b^{-k}$ is regarded as 0 in the case of $m = 1$, in the instructions.

At the beginning of stage m, assume that $f(1), f(2), f(3), \ldots, f(m-1)$ are calculated already. We approximate the real $\beta - \sum_{k=1}^{m-1} f(k)b^{-k}$ and the $\lceil b \rceil - 1$ reals

$$b^{-m}, 2b^{-m}, \ldots, (\lceil b \rceil - 2)b^{-m}, (\lceil b \rceil - 1)b^{-m}$$

by rationals with increasing precision. During the approximation, if we find $l \in \{0, 1, 2, \ldots, \lceil b \rceil - 1\}$ such that

$$lb^{-m} < \beta - \sum_{k=1}^{m-1} f(k)b^{-k} < (l+1)b^{-m}, \tag{7.9}$$

then we set $f(m) := l$ and begin stage $m + 1$.

We can check that our recursion works properly, as follows: First, since $0 < \beta < 1 \leq \lceil b \rceil b^{-1}$, we see that $0 < \beta - \sum_{k=1}^{m-1} f(k)b^{-k} < \lceil b \rceil b^{-m}$ at the beginning of stage $m = 1$. Thus, in general, we assume that $0 < \beta - \sum_{k=1}^{m-1} f(k)b^{-k} < \lceil b \rceil b^{-m}$ at the beginning of stage m. Then, since β and b are computable and Property 7.2 does not hold, we can eventually find $l \in \{0, 1, 2, \ldots, \lceil b \rceil - 1\}$ which satisfies (7.9). Since $b^{-m} \leq \lceil b \rceil b^{-(m+1)}$, we have $0 < \beta - \sum_{k=1}^{m} f(k)b^{-k} < \lceil b \rceil b^{-(m+1)}$ at the beginning of stage $m + 1$.

Thus, Property 7.1 holds in any case. Now, we choose any particular $L \in \mathbb{N}$ with $2^L \geq \lceil b \rceil - 1$. It follows that

$$\sum_{k=1}^{\infty} f(k)2^{-(k+L)} \leq \sum_{k=1}^{\infty} (\lceil b \rceil - 1)2^{-(k+L)} \leq 1. \tag{7.10}$$

First, we consider the case where $f(k) > 0$ for infinitely many $k \in \mathbb{N}^+$. In this case, it follows that there exists a total recursive function $g_1: \mathbb{N} \to \mathbb{N}$ such that $g_1(n) > L$ for all $n \in \mathbb{N}$ and $\#\{n \in \mathbb{N} \mid g_1(n) = k + L\} = f(k)$ for all $k \in \mathbb{N}^+$. We also consider a total recursive function $g_2: \mathbb{N} \to \{0, 1\}^*$ with $g_2(n) = \lambda$. Then, since $\sum_{n=0}^{\infty} 2^{-g_1(n)} \leq 1$ due to (7.10), it follows from Theorem 2.3 that there exists a prefix-free machine M such that $\#\{p \in \{0, 1\}^* \mid |p| = l \ \& \ M(p) = s\} = \#\{n \in \mathbb{N} \mid g_1(n) = l \ \& \ g_2(n) = s\}$ for every $l \in \mathbb{N}$ and $s \in \{0, 1\}^*$. Thus, we have that

(i) $|p| \geq 1 + L$ for every $p \in \operatorname{dom} M$, and
(ii) $\#\{p \in \{0, 1\}^* \mid |p| = k + L \ \& \ p \in \operatorname{dom} M\} = f(k)$ for every $k \in \mathbb{N}^+$.

On the other hand, in the case where $f(k) = 0$ for all but finitely many $k \in \mathbb{N}^+$, based on (7.10) we can more easily show that there exists a prefix-free machine M which satisfies the conditions (i) and (ii) above.

We then define a partial function $V: \{0, 1\}^* \to \{0, 1\}^*$ by the conditions that (i) $\operatorname{dom} V = \{0p \mid p \in \operatorname{dom} U\} \cup \{1p \mid p \in \operatorname{dom} M\}$, (ii) $V(0p) = U(p)$ for all $p \in \operatorname{dom} U$, and (iii) $V(1p) = M(p)$ for all $p \in \operatorname{dom} M$. Since $\operatorname{dom} V$ is a prefix-free set, it follows that V is a prefix-free machine. It is then easy to check that $K_V(s) \leq K_U(s) + 1$ for every $s \in \{0, 1\}^*$. Therefore, since U is an optimal prefix-free machine, V is also optimal. On the other hand, due to Property 7.1, we see that

$$
\begin{aligned}
Z_V(T_1) &= \sum_{p \in \operatorname{dom} U} 2^{-(|p|+1)/T_1} + \sum_{p \in \operatorname{dom} M} 2^{-(|p|+1)/T_1} \\
&= 2^{-1/T_1} Z_U(T_1) + 2^{-(L+1)/T_1} \sum_{k=1}^{\infty} f(k) 2^{-k/T_1} \qquad (7.11) \\
&= 2^{-1/T_1} Z_U(T_1) + 2^{-(L+1)/T_1} \beta.
\end{aligned}
$$

Since T_1 is computable with $0 < T_1 < 1$, it follows from Theorem 7.4 that there exists a T_1-convergent computable, increasing sequence $\{w_n\}$ of rationals which converges to $Z_U(T_1)$. Then, since T_1 is computable, it is easy to show that there exists a computable, increasing sequence $\{a_n\}$ of rationals such that

$$
\eta w_n + \xi c_n < a_n < \eta w_{n+1} + \xi c_{n+1}
$$

for all $n \in \mathbb{N}$, where $\eta := 2^{-1/T_1}$ and $\xi := 2^{-(L+1)/T_1} r$. Obviously, by (7.11) we have $\lim_{n \to \infty} a_n = 2^{-1/T_1} Z_U(T_1) + 2^{-(L+1)/T_1} r \gamma = Z_V(T_1)$. Using the inequality $(x + y)^t \leq x^t + y^t$ for reals $x, y > 0$ and $t \in (0, 1]$, we have

$$
\begin{aligned}
(a_{n+1} - a_n)^T &< \left[(\eta w_{n+2} + \xi c_{n+2}) - (\eta w_n + \xi c_n)\right]^T \\
&\leq \eta^T (w_{n+2} - w_n)^T + \xi^T (c_{n+2} - c_n)^T \\
&\leq \eta^T (w_{n+2} - w_{n+1})^T + \eta^T (w_{n+1} - w_n)^T \\
&\quad + \xi^T (c_{n+2} - c_{n+1})^T + \xi^T (c_{n+1} - c_n)^T
\end{aligned}
$$

for each $T \in (0, 1]$. Thus, for each $T \in (T_2, 1]$, since both $\{w_n\}$ and $\{c_n\}$ are T-convergent, $\{a_n\}$ is also T-convergent. On the other hand, using the same inequality, we have

$$
\begin{aligned}
\xi^T (c_{n+2} - c_{n+1})^T &< \left[\eta(w_{n+2} - w_{n+1}) + \xi(c_{n+2} - c_{n+1}) \right]^T \\
&= \left[(\eta w_{n+2} + \xi c_{n+2}) - (\eta w_{n+1} + \xi c_{n+1}) \right]^T \\
&< (a_{n+2} - a_n)^T \\
&\leq (a_{n+2} - a_{n+1})^T + (a_{n+1} - a_n)^T
\end{aligned}
$$

for each $T \in (0, 1]$. Thus, for each $T \in (0, T_2]$, since $\{c_n\}$ is not T-convergent, it is easy to see that $\{a_n\}$ is not T-convergent also. This completes the proof. □

7.7 Future Direction

In this chapter, we have generalized the equivalent characterizations of randomness for an r.e. real over the notion of partial randomness, so that the generalized characterizations are all equivalent to the weak Chaitin T-randomness. As a stronger notion of the partial randomness of a real, in the preceding chapters we have introduced and studied the notion of the Chaitin T-randomness. Thus, future work may aim at modifying our equivalent characterizations of partial randomness given by Theorem 7.5, so that they become equivalent to the Chaitin T-randomness.

In the next chapter, we reveal further the properties of the thermodynamic quantities of AIT, related to the notion of T-convergence, based on Theorem 7.5.

Chapter 8
Computation-Theoretic Clarification of the Phase Transition at Temperature $T = 1$

8.1 Computational Complexity-Theoretic Treatment in Computability Theory

The notion of weak truth-table reducibility plays an important role in recursion theory (see e.g. Nies [27] and Downey and Hirschfeldt [14]). For any sets $A, B \subset \mathbb{N}$, we say that A *is weak truth-table reducible to* B, denoted $A \leq_{wtt} B$, if there exist an oracle Turing machine M and a total recursive function $g: \mathbb{N} \to \mathbb{N}$ such that (i) A is Turing reducible to B via M and (ii) on every input $n \in \mathbb{N}$, the machine M only queries natural numbers at most $g(n)$. In this chapter, we introduce an elaboration of this notion, where the total recursive bound g on the *use* of the reduction is explicitly specified. In doing so, in particular *we try to follow the fashion in which computational complexity theory is developed, while staying in computability theory*. We apply the elaboration to the sets which appear in AIT. The elaboration, called *reducibility in query size f*, is introduced as follows.

Definition 8.1 (*Reducibility in Query Size f, Tadaki [51]*) Let $f: \mathbb{N} \to \mathbb{N}$, and let $A, B \subset \{0, 1\}^*$. We say that A *is reducible to B in query size f* if there exists an oracle Turing machine M such that

(i) A is Turing reducible to B via M, and
(ii) on every input $x \in \{0, 1\}^*$, the machine M only queries strings of length at most $f(|x|)$. ☐

For any fixed sets A and B, the above definition allows us to consider the notion of *asymptotic behavior* regulated by the function f, which bounds the use of the reduction, i.e., which imposes the restriction on the use of the computational resource (i.e., the oracle B). Thus, by the above definition, *even in the context of computability theory, we can deal with the notion of asymptotic behavior in a manner like in computational complexity theory*. Recall here that the notion of input size plays

This chapter is a rearrangement of Tadaki [46, 51].

a crucial role in computational complexity theory since computational complexity such as time complexity and space complexity is measured based on it. This is also true in AIT since the program-size complexity is measured based on input size. Thus, in Definition 8.1 we consider a reduction between subsets of $\{0, 1\}^*$ and not a reduction between subsets of $\mathbb{N}$ as in the original weak truth-table reducibility. Moreover, in Definition 8.1 we require the bound $f(|x|)$ to depend only on input size $|x|$ as in computational complexity theory, and not on input x itself as in the original weak truth-table reducibility. *We pursue a formal correspondence to computational complexity theory in this manner*, while *staying in computability theory*.

In this chapter we demonstrate the power of the notion of reducibility in query size f in the context of AIT, especially based on Chaitin's halting probability Ω and its generalization $Z(T)$ over the notion of partial randomness. Recall that the first n bits of the base-two expansion of Ω (i.e., Ω_U) solve the halting problem of the optimal prefix-free machine U for all binary inputs of length at most n. Using this property, Chaitin showed Ω to be a random real, as stated in Theorem 2.5. Recall that dom U is the set of all halting inputs for U. Calude and Nies [3], in essence, showed the following theorem on the relation between the base-two expansion of Ω and the halting problem dom U.

Theorem 8.1 (Calude and Nies [3]) *The real Ω and the set* dom U *are weak truth-table equivalent. Namely,* $\Omega \leq_{wtt}$ dom U *and* dom $U \leq_{wtt} \Omega$. $\qquad\square$

Similarly, we can show Theorem 8.2 below for the partition function $Z(T)$ of AIT, defined in Definition 4.1. This theorem follows immediately from stronger results, Theorems 8.19 and 8.20, which are two of the main results of this chapter.

Theorem 8.2 *Suppose that T is a computable real with $0 < T < 1$. Then $Z(T)$ and* dom U *are weak truth-table equivalent.* $\qquad\square$

First, recall that $\Omega = Z(1)$. When comparing Theorem 8.1 and Theorem 8.2, we see that there is no difference between $T = 1$ and $T < 1$ with respect to the weak truth-table equivalence between $Z(T)$ and dom U. In this chapter, however, we show that there is a critical difference between $T = 1$ and $T < 1$ in the relation between $Z(T)$ and dom U from the point of view of the reducibility in query size f. Based on the notion of reducibility in query size f, we introduce the notions of *unidirectionality* and *bidirectionality* between two sets A and B in this chapter. These notions enable us to investigate the *relative computational power* between A and B.

Theorems 8.4 and 8.5 below are two of the main results of this chapter. Theorem 8.4 gives a succinct equivalent characterization of f for which Ω is reducible to dom U in query size f and reversely Theorem 8.5 gives a succinct equivalent characterization of f for which dom U is reducible to Ω in query size f, both in a general setting. Based on them, we show in Theorem 8.6 below that the computation from Ω to dom U is *unidirectional* and the computation from dom U to Ω is also *unidirectional*. On the other hand, Theorems 8.19 and 8.20 below are also two of the main results of this chapter. Theorem 8.19 gives a succinct equivalent characterization of f for which $Z(T)$ is reducible to dom U in query size f and reversely

Theorem 8.20 gives a succinct equivalent characterization of f for which dom U is reducible to $Z(T)$ in query size f, both in a general setting, in the case where T is a computable real with $0 < T < 1$. Based on them, we show in Theorem 8.21 below that the computations between $Z(T)$ and dom U are *bidirectional* if T is a computable real with $0 < T < 1$. In this way the notion of reducibility in query size f can reveal a critical difference of the behavior of $Z(T)$ between $T = 1$ and $T < 1$, which cannot be captured by the original notion of weak truth-table reducibility.

Recall that the weak truth-table reducibility is defined for two subsets of $\mathbb{N}$. Thus, when we apply the notion of weak truth-table reducibility to a real α, we regard α as a subset of $\mathbb{N}$ whose characteristic sequence equals to the base-two expansion of α (i.e., Binary (α)). In fact, in Theorem 8.1, the real Ω is regarded as a subset of $\mathbb{N}$ in this manner. On the other hand, the notion of reducibility in query size f is defined for two subsets of $\{0, 1\}^*$. Thus, when we apply this notion to a real, we have to somehow regard it as a subset of $\{0, 1\}^*$. We do this by using the following notion of *prefixes* of a real.

Definition 8.2 *(Prefixes of Real)* For each $\alpha \in \mathbb{R}$, the *prefixes* $\mathrm{Pf}(\alpha)$ of α is the subset of $\{0, 1\}^*$ defined by $\mathrm{Pf}(\alpha) = \{\alpha\lceil_n \mid n \in \mathbb{N}\}$. $\qquad\qquad\square$

The notion of prefixes of a real is already a natural notion in AIT. For example, the weak Chaitin randomness of a real α can be rephrased as that there exists $d \in \mathbb{N}$ such that, for every $x \in \mathrm{Pf}(\alpha)$, it holds that $|x| \leq K(x) + d$. The notion of prefixes of a real helps us see the aforementioned unidirectionality and bidirectionality.

In Chap. 4, we have seen that the values of all the thermodynamic quantities diverge when the temperature T exceeds 1. This phenomenon might be regarded as some sort of phase transition in statistical mechanics. Thus, in this chapter we reveal a *new aspect* of the phase transition by showing the critical difference of the behavior of $Z(T)$ between $T = 1$ and $T < 1$ in terms of the notion of reducibility in query size f.

Tadaki [46] considered an elaboration of weak truth-table equivalence between Ω and dom U and *already* revealed the *unidirectionality* between them. In this chapter, based on the notion of reducibility in query size f, we rephrase the results of Tadaki [46] *with a thorough emphasis on a formal correspondence to computational complexity theory*.

This chapter is organized as follows. In Sect. 8.2 we introduce some convention on the computation of a deterministic Turing machine which computes a prefix-free machine, and then show a result based on it. In Sect. 8.3 we investigate simple properties of the notion of reducibility in query size f and introduce the notions of unidirectionality and bidirectionality between two sets based on it. We then show in Sect. 8.4 the unidirectionality between Ω and dom U in a general setting. In Sect. 8.5 we present theorems which play a crucial role in establishing the bidirectionality in Sect. 8.6. Based on them, we show in Sect. 8.6 the bidirectionality between $Z(T)$ and dom U with a computable real $T \in (0, 1)$ in a general setting. We conclude this chapter with the remarks on the origin of the phase transition of the behavior of $Z(T)$ between $T = 1$ and $T < 1$ in Sect. 8.7.

In this chapter, for any subset S of $\{0, 1\}^*$ and any $n \in \mathbb{N}$, we denote by $S\!\upharpoonright_n$ the set $\{s \in S \mid |s| \leq n\}$. In addition, in this chapter we often refer to the inequality (8.1) below, which is Theorem 3.1 (g) of Chaitin [9]. Recall that in Sect. 2.3 we have defined $K(s, t)$ as $K(b(s, t))$ for any $s, t \in \{0, 1\}^*$, where b is the 'standard' bijective total recursive function from $\{0, 1\}^* \times \{0, 1\}^*$ to $\{0, 1\}^*$. Consider a prefix-free machine F such that, for every $p, v \in \{0, 1\}^*$, $F(p) = v$ if and only if there exist $q, r \in \operatorname{dom} U$ with the properties that $p = qr$ and $v = b(U(q), U(r))$. Then, applying (2.4) to F, we have that

$$K(s, t) \leq K(s) + K(t) + O(1) \tag{8.1}$$

for all $s, t \in \{0, 1\}^*$.

8.2　Turing Machine Computing a Prefix-Free Machine

Let M be a deterministic Turing machine with the input and output alphabet $\{0, 1\}$, and let F be a prefix-free machine. We say that M *computes* F if the following holds: for every $p \in \{0, 1\}^*$, when M starts with the input p, (i) M halts and outputs $F(p)$ if $p \in \operatorname{dom} F$; (ii) M does not halt forever otherwise. We use this convention on the computation of a prefix-free machine by a deterministic Turing machine throughout the rest of this chapter. Thus, we exclude the possibility that there is $p \in \{0, 1\}^*$ such that, when M starts with the input p, M halts but $p \notin \operatorname{dom} F$. For any $p \in \{0, 1\}^*$, we denote the running time of M on the input p by $T_M(p)$ (may be ∞). Thus, $T_M(p) \in \mathbb{N}$ for every $p \in \operatorname{dom} F$ if M computes F.

We define $L_M := \min\{|p| \mid p \in \{0, 1\}^* \,\&\, M \text{ halts on input } p\}$ (may be ∞). For any $n \geq L_M$, we define I_M^n as the set of all halting inputs p for M with $|p| \leq n$ which take longest to halt in the computation of M, i.e., as the set

$$\{p \in \{0, 1\}^* \mid |p| \leq n \,\&\, T_M(p) = T_M^n\}$$

where T_M^n is the maximum running time of M on all halting inputs of length at most n. Tadaki [46] slightly strengthened the result presented in Chaitin [12] to obtain Theorem 8.3 below (see Note in Sect. 8.1 of Chaitin [12]).

Theorem 8.3 (Chaitin [12] and Tadaki [46]) *Let V be an optimal prefix-free machine, and let M be a deterministic Turing machine which computes V. Then $n = K(n, p) + O(1) = K(p) + O(1)$ for all (n, p) with $n \geq L_M$ and $p \in I_M^n$.* □

Theorem 8.3 is proved as follows. We first need Lemma 8.1 below to prove it.

Lemma 8.1 *Let V be an optimal prefix-free machine, and let M be a deterministic Turing machine which computes V. Then there exists $d \in \mathbb{N}$ such that, for every $p \in \operatorname{dom} V$, there exists $q \in \operatorname{dom} V$ for which $|q| \leq |p| + d$ and $T_M(q) > T_M(p)$.*

Proof Consider a prefix-free machine F such that (i) dom $F =$ dom V and (ii) $F(p) = 1^{2|p|+T_M(p)+1}$ for every $p \in$ dom V. It is easy to see that such a prefix-free machine F exists. Then, since V is an optimal prefix-free machine, from the definition of an optimal prefix-free machine there exists $d_1 \in \mathbb{N}$ with the following property; if $p \in$ dom F, then there is q for which $V(q) = F(p)$ and $|q| \le |p| + d_1$.

Thus, for each $p \in$ dom V with $|p| \ge d_1$, there is q for which $V(q) = F(p)$ and $|q| \le |p| + d_1$. It follows that

$$|V(q)| = 2|p| + T_M(p) + 1 > |p| + d_1 + T_M(p) \ge |q| + T_M(p). \tag{8.2}$$

Note that exactly $|q|$ cells on the tapes of M have the symbols 0 or 1 in the initial configuration of M with the input q, while at least $|V(q)|$ cells on the tape of M, where the output is put, have the symbols 0 or 1 in the resulting final configuration of M. Since M can write at most one 0 or 1 on the tape, where the output is put, every one step of its computation, the running time $T_M(q)$ of M on the input q is bounded to the below by the difference $|V(q)| - |q|$. Thus, by (8.2), we have $T_M(q) > T_M(p)$.

On the other hand, since dom V is not a recursive set, the function T_n^M of n with $n \ge L_M$ is not bounded to the above. Therefore, there exists $r_0 \in$ dom V such that, for every $p \in$ dom F with $|p| < d_1$, it holds that $T_M(r_0) > T_M(p)$. Then, by setting $d_2 := |r_0|$, we have that $|r_0| \le |p| + d_2$ for every $p \in$ dom F with $|p| < d_1$, obviously.

Thus, by setting $d := \max\{d_1, d_2\}$ we see that, for every $p \in$ dom V, there is $q \in$ dom V for which $|q| \le |p| + d$ and $T_M(q) > T_M(p)$. $\qquad \square$

Then the proof of Theorem 8.3 is given as follows.

Proof (of Theorem 8.3) By considering the following procedure, we first show that $n \le K(n, p) + O(1)$ for all (n, p) with $n \ge L_M$ and $p \in I_M^n$.

Given (n, p) with $n \ge L_M$ and $p \in I_M^n$, one first calculates the finite set dom $V\!\restriction_n$ by simulating the computation of M with the input q until at most $T_M(p)$ steps, for each $q \in \{0, 1\}^*$ with $|q| \le n$. Then, by calculating the set $\{V(q) \mid q \in$ dom $V\!\restriction_n\}$ and picking any particular finite binary string which is not in this set, one can obtain an $s \in \{0, 1\}^*$ such that $n < K_V(s)$.

Hence, there exists a partial recursive function $\Psi : \mathbb{N}^+ \times \{0, 1\}^* \to \{0, 1\}^*$ such that $n < K_V(\Psi(n, p))$ for all (n, p) with $n \ge L_M$ and $p \in I_M^n$. It follows from the optimality of V and U that $n < K(n, p) + O(1)$ for all (n, p) with $n \ge L_M$ and $p \in I_M^n$.

We next show that $K(n, p) \le K(p) + O(1)$ for all (n, p) with $n \ge L_M$ and $p \in I_M^n$. From Lemma 8.1 we first note that there exists $d \in \mathbb{N}$ such that, for every $p \in$ dom V, there exists $q \in$ dom V for which $|q| \le |p| + d$ and $T_M(q) > T_M(p)$. Then, for each (n, p) with $n \ge L_M$ and $p \in I_M^n$, we see that $|p| \le n$ due to the definition of I_M^n, and also there exists $q \in$ dom V for which $|q| \le |p| + d$ and $T_M(q) > T_M(p)$. Note here that $T_M(q) > T_M^n$ due to the definition of I_M^n again, and therefore $|q| > n$ due to the definition of T_M^n. Thus $|p| \le n < |p| + d$ and $d \ge 1$. Hence, given p such that $n \ge L_M$ and $p \in I_M^n$, one only needs to specify one of d possibilities of n in order to calculate n.

Thus, there exists a partial recursive function $\Phi: \{0, 1\}^* \times \mathbb{N}^+ \to \mathbb{N}^+ \times \{0, 1\}^*$ such that, for every (n, p) with $n \geq L_M$ and $p \in I_M^n$, there exists $k \in \mathbb{N}^+$ with the properties that $1 \leq k \leq d$ and $\Phi(p, k) = (n, p)$. It follows from (2.8) that there exists $c \in \mathbb{N}$ such that, for every (n, p) with $n \geq L_M$ and $p \in I_M^n$, there exists $k \in \mathbb{N}^+$ with the properties that $1 \leq k \leq d$ and $K(n, p) \leq K(p, k) + c$. Thus, using (8.1) we have that

$$K(n, p) \leq K(p) + \max\{K(k) \mid k \in \mathbb{N}^+ \ \& \ 1 \leq k \leq d\} + O(1)$$

for all (n, p) with $n \geq L_M$ and $p \in I_M^n$.

Finally, we show that $K(p) \leq n + O(1)$ for all (n, p) with $n \geq L_M$ and $p \in I_M^n$. Let us consider a prefix-free machine F such that (i) dom $F =$ dom V and (ii) $F(p) = p$ for every $p \in$ dom V. Obviously, such a prefix-free machine F exists. Then it follows from (2.4) that $K(p) \leq |p| + O(1)$ for all $p \in$ dom V. For each (n, p) with $n \geq L_M$ and $p \in I_M^n$, we see from the definition of I_M^n that $p \in$ dom V and $|p| \leq n$, and therefore we have that $K(p) \leq |p| + O(1) \leq n + O(1)$. This completes the proof.

□

8.3 Reducibility in Query Size f

In this section we investigate some properties of the notion of reducibility in query size f and introduce the notions of *unidirectionality* and *bidirectionality* between two sets.

First note that, for every set $A \subset \{0, 1\}^*$, A is reducible to A in query size n, where "n" denotes the identity function $I: \mathbb{N} \to \mathbb{N}$ with $I(n) = n$. In this manner we follow the notation in computational complexity theory. The following are simple observations on the notion of reducibility in query size f.

Proposition 8.1 *Let* $f: \mathbb{N} \to \mathbb{N}$ *and* $g: \mathbb{N} \to \mathbb{N}$, *and let* $A, B, C \subset \{0, 1\}^*$.

  (i) *If A is reducible to B in query size f and B is reducible to C in query size g, then A is reducible to C in query size $g \circ f$.*

 (ii) *Suppose that $f(n) \leq g(n)$ for every $n \in \mathbb{N}$. If A is reducible to B in query size f then A is reducible to B in query size g.*

(iii) *Suppose that A is reducible to B in query size f. If A is not recursive then f is unbounded.*   □

The following proposition is a restatement of the fact that, for every optimal prefix-free machine V, the first n bits of the base-two expansion of Ω_V solve the halting problem of V for inputs of length at most n.

Proposition 8.2 *Let V be an optimal prefix-free machine. Then* dom V *is reducible to* $\mathrm{Pf}(\Omega_V)$ *in query size n.*   □

Definition 8.3 An *order function* is a non-decreasing total recursive function $f: \mathbb{N} \to \mathbb{N}$ such that $\lim_{n \to \infty} f(n) = \infty$.   □

Let f be an order function. Intuitively, the notion of the reduction of A to B in query size f is equivalent to that, for every $n \in \mathbb{N}$, if n and $B \restriction_{f(n)}$ are given, then the finite set $A \restriction_n$ can be calculated. With this view in mind, we introduce the notions of unidirectionality and bidirectionality between two sets as follows.

Definition 8.4 *(Tadaki [51])* Let $A, B \subset \{0, 1\}^*$. We say that *the computation from A to B is unidirectional* if the following holds: For every order functions f and g, if B is reducible to A in query size f and A is reducible to B in query size g then the function $g(f(n)) - n$ of $n \in \mathbb{N}$ is unbounded. We say that *the computations between A and B are bidirectional* if the computation from A to B is not unidirectional and the computation from B to A is not unidirectional. $\qquad\square$

The notion of unidirectionality of the computation from A to B in the above definition is, in essence, interpreted as follows: No matter how an order function f is chosen so as to satisfy that the finite set $B \restriction_n$ can be calculated from n and $A \restriction_{f(n)}$, the finite set $A \restriction_{f(n)}$ cannot be calculated from n and $B \restriction_{n+O(1)}$.

8.4 Unidirectionality

In this section we show the *unidirectionality* between Ω_U and dom U in a *general setting*. Theorems 8.4 and 8.5 below are two of the main results of this chapter.

Theorem 8.4 (Elaboration of $\Omega_U \leq_{wtt}$ dom U) *Let V and W be optimal prefix-free machines, and let f be an order function. Then the following two conditions are equivalent:*

(i) $\mathrm{Pf}(\Omega_V)$ *is reducible to* dom W *in query size* $f(n) + O(1)$.
(ii) $\sum_{n=0}^{\infty} 2^{n-f(n)} < \infty$. $\qquad\square$

Theorem 8.4 is proved in Sect. 8.4.1 below. Theorem 8.4 corresponds to Theorem 4 of Tadaki [46], and is proved by modifying the proof of Theorem 4 of Tadaki [46]. Let V and W be optimal prefix-free machines. The implication (ii) $\Rightarrow$ (i) of Theorem 8.4 results in, for example, that $\mathrm{Pf}(\Omega_V)$ is reducible to dom W in query size $n + \lfloor (1 + \varepsilon) \log_2 n \rfloor + O(1)$ for every real $\varepsilon > 0$. On the other hand, the implication (i) $\Rightarrow$ (ii) of Theorem 8.4 results in, for example, that $\mathrm{Pf}(\Omega_V)$ is not reducible to dom W in query size $n + \lfloor \log_2 n \rfloor + O(1)$ and therefore, in particular, $\mathrm{Pf}(\Omega_V)$ is not reducible to dom W in query size $n + O(1)$.

Theorem 8.5 (Elaboration of dom $U \leq_{wtt} \Omega_U$) *Let V and W be optimal prefix-free machines, and let f be an order function. Then the following two conditions are equivalent:*

(i) dom W *is reducible to* $\mathrm{Pf}(\Omega_V)$ *in query size* $f(n) + O(1)$.
(ii) $n \leq f(n) + O(1)$. $\qquad\square$

Theorem 8.5 is proved in Sect. 8.4.2 below. Theorem 8.5 corresponds to Theorem 11 of Tadaki [46], and is proved by modifying the proof of Theorem 11 of Tadaki [46]. The implication (ii) $\Rightarrow$ (i) of Theorem 8.5 results in that, for every optimal prefix-free machines V and W, dom W is reducible to $\mathrm{Pf}(\Omega_V)$ in query size $n + O(1)$. On the other hand, the implication (i) $\Rightarrow$ (ii) of Theorem 8.5 says that this upper bound "$n + O(1)$" of the query size is, in essence, tight.

Theorem 8.6 *Let V and W be optimal prefix-free machines. Then the computation from $\mathrm{Pf}(\Omega_V)$ to dom W is unidirectional and the computation from dom W to $\mathrm{Pf}(\Omega_V)$ is also unidirectional.*

Proof Let V and W be optimal prefix-free machines. For arbitrary order functions f and g, assume that dom W is reducible to $\mathrm{Pf}(\Omega_V)$ in query size f and $\mathrm{Pf}(\Omega_V)$ is reducible to dom W in query size g. It follows from the implication (i) $\Rightarrow$ (ii) of Theorem 8.5 that there exists $c \in \mathbb{N}$ for which $n \leq f(n) + c$ for all $n \in \mathbb{N}$. On the other hand, it follows from the implication (i) $\Rightarrow$ (ii) of Theorem 8.4 that $\sum_{n=0}^{\infty} 2^{n-g(n)} < \infty$ and therefore $\lim_{n\to\infty} g(n) - n = \infty$. Since g is an order function, we have $g(f(n)) - n \geq g(n - c) - (n - c) - c$ for all $n \geq c$. Thus, the computation from $\mathrm{Pf}(\Omega_V)$ to dom W is unidirectional. On the other hand, we have $f(g(n)) - n \geq g(n) - n - c$ for all $n \in \mathbb{N}$. Thus, the computation from dom W to $\mathrm{Pf}(\Omega_V)$ is unidirectional. $\square$

8.4.1 The Proof of Theorem 8.4

Theorem 8.4 follows from Theorems 8.9 and 8.11 below, and the fact that Ω_V is a weakly Chaitin random r.e. real for every optimal prefix-free machine V. We first prove Theorem 8.9 using Theorems 2.2 and 8.8 below.

To begin with, we prove Theorem 8.7 below. Theorem 8.8 is a restatement of it using the notion of reducibility in query size f.

Theorem 8.7 (Tadaki [46]) *Let V be an optimal prefix-free machine. Then, for every prefix-free machine F, there exists $d \in \mathbb{N}$ such that, for every $p \in \{0, 1\}^*$, if p and the list of all halting inputs for V of length at most $|p| + d$, i.e., the finite set $\mathrm{dom}\, V\!\upharpoonright_{|p|+d}$, are given, then the halting problem of the input p for F can be solved.*

Proof Let M be a deterministic Turing machine which computes a prefix-free machine F. We consider a prefix-free machine G such that (i) dom $G = $ dom F and (ii) $G(p) = T_M(p)$ for every $p \in$ dom F. Recall here that we identify $\{0, 1\}^*$ with $\mathbb{N}$. It is easy to see that such a prefix-free machine G exists. Then, since V is an optimal prefix-free machine, from the definition of optimality there exists $d \in \mathbb{N}$ with the following property; if $p \in$ dom G, then there is $q \in$ dom V for which $V(q) = G(p)$ and $|q| \leq |p| + d$.

Given $p \in \{0, 1\}^*$ and the list $\{q_1, \ldots, q_L\}$ of all halting inputs for V of length at most $|p| + d$, one first calculates a finite set $S := \{V(q_i) \mid i = 1, \ldots, L\}$, and then calculates $T_{max} := \max S$ where S is regarded as a subset of $\mathbb{N}$. One then simulates the computation of M with the input p until at most the time step T_{max}. In the simulation, if M halts until at most the time step T_{max}, one knows that $p \in \text{dom } F$. On the other hand, note that if $p \in \text{dom } F$ then there is $q \in \text{dom } V$ such that $V(q) = T_M(p)$ and $|q| \leq |p| + d$, and therefore $q \in \{q_1, \ldots, q_L\}$ and $T_M(p) \leq T_{max}$. Thus, in the simulation, if M does not yet halt at the time step T_{max}, one knows that M does not halt forever and therefore $p \notin \text{dom } F$. □

Theorem 8.8 *Let V be an optimal prefix-free machine. Then, for every prefix-free machine F, there exists $d \in \mathbb{N}$ such that* dom F *is reducible to* dom V *in query size $n + d$.* □

Theorem 8.9 *Let α be an r.e. real, and let V be an optimal prefix-free machine. For every total recursive function $f: \mathbb{N} \to \mathbb{N}$, if $\sum_{n=0}^{\infty} 2^{n-f(n)} < \infty$, then there exists $c \in \mathbb{N}$ such that* $\text{Pf}(\alpha)$ *is reducible to* dom V *in query size $f(n) + c$.*

Proof Let α be an r.e. real, and let V be an optimal prefix-free machine. For an arbitrary total recursive function $f: \mathbb{N} \to \mathbb{N}$, assume that $\sum_{n=0}^{\infty} 2^{n-f(n)} < \infty$. In the case of $\alpha \in \mathbb{Q}$, the result is obvious. Thus, in what follows, we assume that $\alpha \notin \mathbb{Q}$ and therefore the infinite binary sequence Binary (α) contains infinitely many ones.

Since $\sum_{n=0}^{\infty} 2^{n-f(n)} < \infty$, there exists $d_0 \in \mathbb{N}$ such that $\sum_{n=0}^{\infty} 2^{n-f(n)-d_0} \leq 1$. Hence, by the Kraft-Chaitin Theorem, i.e., Theorem 2.2, there exists a total recursive function $g: \mathbb{N} \to \{0, 1\}^*$ such that (i) the function g is an injection, (ii) the set $\{g(n) \mid n \in \mathbb{N}\}$ is prefix-free, and (iii) $|g(n)| = f(n) - n + d_0$ for all $n \in \mathbb{N}$. On the other hand, since α is r.e., there exists a total recursive function $h: \mathbb{N} \to \mathbb{Q}$ such that $h(k) \leq \alpha$ for all $k \in \mathbb{N}$ and $\lim_{k \to \infty} h(k) = \alpha$.

Now, let us consider a prefix-free machine F such that, for every $p, v \in \{0, 1\}^*$, $F(p) = v$ if and only if there exist $n \in \mathbb{N}$ and $s \in \{0, 1\}^*$ with the properties that (i) $p = g(n)s$, (ii) $|s| = n$, (iii) $0.s < h(k) - \lfloor \alpha \rfloor$ for some $k \in \mathbb{N}$, and (iv) $v = \lambda$. Note that such a prefix-free machine F exists. Since Binary (α) contains infinitely many ones, it follows that, for every $n \in \mathbb{N}$ and $s \in \{0, 1\}^n$,

$$g(n)s \in \text{dom } F \text{ if and only if } s \leq \alpha\lceil_n, \tag{8.3}$$

where s and $\alpha\lceil_n$ are regarded as a dyadic integer. Then, by the following procedure, we see that $\text{Pf}(\alpha)$ is reducible to dom F in query size $f(n) + d_0$.

Given $t \in \{0, 1\}^*$, based on the equivalence (8.3), one determines $\alpha\lceil_n$ by putting the queries $g(n)s$ to the oracle dom F for all $s \in \{0, 1\}^n$, where $n = |t|$. Note here that all the queries are of length $f(n) + d_0$, since $|g(n)| = f(n) - n + d_0$. One then accepts if $t = \alpha\lceil_n$ and rejects otherwise.

On the other hand, by Theorem 8.8, there exists $d \in \mathbb{N}$ such that dom F is reducible to dom V in query size $n + d$. Thus, it follows from Proposition 8.1 (i) that $\text{Pf}(\alpha)$ is reducible to dom V in query size $f(n) + d_0 + d$, as desired. □

We next prove Theorem 8.11 below using Theorem 8.3 and the Ample Excess Lemma below.

Theorem 8.10 (Ample Excess Lemma, Miller and Yu [26]) *For every $\alpha \in \mathbb{R}$, α is weakly Chaitin random if and only if $\sum_{n=1}^{\infty} 2^{n-K(\alpha\upharpoonright n)} < \infty$.* $\square$

Theorem 8.11 *Let α be a real which is weakly Chaitin random, and let V be an optimal prefix-free machine. For every order function f, if $\mathrm{Pf}(\alpha)$ is reducible to $\mathrm{dom}\, V$ in query size f then $\sum_{n=0}^{\infty} 2^{n-f(n)} < \infty$.*

Proof Let α be a real which is weakly Chaitin random, and let V be an optimal prefix-free machine. For an arbitrary order function f, assume that $\mathrm{Pf}(\alpha)$ is reducible to $\mathrm{dom}\, V$ in query size f. Since f is an order function, a set

$$S_f := \{n \in \mathbb{N} \mid f(n) < f(n+1)\}$$

is an infinite recursive set. Therefore there exists an increasing total recursive function $h \colon \mathbb{N} \to \mathbb{N}$ such that $h(\mathbb{N}) = S_f$. It is then easy to see that $f(n) = f(h(k+1))$ for every k and n with $h(k) < n \le h(k+1)$. Thus, for each $k \ge 1$, we see that

$$
\sum_{n=h(0)+1}^{h(k)} 2^{n-f(n)} = \sum_{j=0}^{k-1} \sum_{n=h(j)+1}^{h(j+1)} 2^{n-f(n)} = \sum_{j=0}^{k-1} 2^{-f(h(j+1))} \sum_{n=h(j)+1}^{h(j+1)} 2^{n}
$$
$$
= \sum_{j=0}^{k-1} 2^{-f(h(j+1))} \left(2^{h(j+1)+1} - 2^{h(j)+1} \right) < 2 \sum_{j=1}^{k} 2^{h(j)-f(h(j))}. \tag{8.4}
$$

On the other hand, let M be a particular deterministic Turing machine which computes V. For each $l \ge L_M$, we choose a particular p_l from I_M^l. Then, given $(n, p_{f(n)})$ with $n \in \mathbb{N}^+$ and $f(n) \ge L_M$, one can calculate the finite set $\mathrm{dom}\, V \upharpoonright_{f(n)}$ by simulating the computation of M with the input q until at most the time step $T_M(p_{f(n)})$, for each $q \in \{0,1\}^*$ with $|q| \le f(n)$. This can be possible because $T_M(p_{f(n)}) = T_M^{f(n)}$ for every $n \in \mathbb{N}^+$ with $f(n) \ge L_M$. Thus, since $\mathrm{Pf}(\alpha)$ is reducible to $\mathrm{dom}\, V$ in query size f by the assumption, we have that there exists a partial recursive function $\Psi \colon \mathbb{N}^+ \times \{0,1\}^* \to \{0,1\}^*$ such that $\Psi(n, p_{f(n)}) = \alpha\upharpoonright_n$ for every $n \in \mathbb{N}^+$ with $f(n) \ge L_M$. It follows from the inequality (2.8) that

$$K(\alpha\upharpoonright_n) \le K(n, p_{f(n)}) + O(1) \tag{8.5}$$

for all $n \in \mathbb{N}^+$ with $f(n) \ge L_M$. Then, since the mapping $\mathbb{N} \ni k \mapsto f(h(k))$ is an increasing total recursive function, it follows from the optimality of U that $K(h(k), s) \le K(f(h(k)), s) + O(1)$ for all $k \in \mathbb{N}$ and $s \in \{0,1\}^*$. Therefore, using (8.5) and Theorem 8.3 we have that

$$K(\alpha\upharpoonright_{h(k)}) \le f(h(k)) + O(1) \tag{8.6}$$

for all $k \in \mathbb{N}^+$. Since α is weakly Chaitin random, using the Ample Excess Lemma, i.e., Theorem 8.10, we have $\sum_{n=1}^{\infty} 2^{n-K(\alpha\upharpoonright n)} < \infty$. Recall that the function h is increasing. Thus, using (8.6) we have

$$\sum_{j=1}^{\infty} 2^{h(j)-f(h(j))} \leq \sum_{j=1}^{\infty} 2^{h(j)-K(\alpha\restriction_{h(j)})+O(1)} \leq \sum_{n=1}^{\infty} 2^{n-K(\alpha\restriction_n)+O(1)} < \infty.$$

It follows from (8.4) that $\lim_{k\to\infty} \sum_{n=h(0)+1}^{h(k)} 2^{n-f(n)} < \infty$. Thus, since $2^{n-f(n)} > 0$ for all $n \in \mathbb{N}$ and $\lim_{k\to\infty} h(k) = \infty$, we have $\sum_{n=0}^{\infty} 2^{n-f(n)} < \infty$, as desired. $\qquad\square$

8.4.2 The Proof of Theorem 8.5

The implication (ii) $\Rightarrow$ (i) of Theorem 8.5 follows immediately from Proposition 8.2 and Proposition 8.1 (ii). On the other hand, the implication (i) $\Rightarrow$ (ii) of Theorem 8.5 is proved as follows.

Proof (of (i) $\Rightarrow$ (ii) of Theorem 8.5) Let V and W be optimal prefix-free machines, and let f be an order function. Suppose that there exists $c \in \mathbb{N}$ such that dom W is reducible to $\mathrm{Pf}(\Omega_V)$ in query size $f(n) + c$. Then, by considering the following procedure, we first see that $n < K(n, \Omega_V\restriction_{f(n)+c}) + O(1)$ for all $n \in \mathbb{N}$.

Given n and $\Omega_V\restriction_{f(n)+c}$, one first calculates the finite set dom $W\restriction_n$. This is possible since dom W is reducible to $\mathrm{Pf}(\Omega_V)$ in query size $f(n) + c$ and $f(k) \leq f(n)$ for all $k \leq n$. Then, by calculating the set $\{ W(p) \mid p \in \text{dom } W\restriction_n \}$ and picking any particular finite binary string which is not in this set, one can obtain an $s \in \{0, 1\}^*$ such that $n < K_W(s)$.

Thus, there exists a partial recursive function $\Psi : \mathbb{N} \times \{0, 1\}^* \to \{0, 1\}^*$ such that $n < K_W(\Psi(n, \Omega_V\restriction_{f(n)+c}))$ for every $n \in \mathbb{N}$. It follows from the optimality of W that

$$n < K(n, \Omega_V\restriction_{f(n)+c}) + O(1) \tag{8.7}$$

for all $n \in \mathbb{N}$.

Now, let us assume contrarily that the function $n - f(n)$ of $n \in \mathbb{N}$ is unbounded. Then, since f is an order function, it is easy to show that there exists a total recursive function $g : \mathbb{N} \to \mathbb{N}$ such that the function $f(g(k))$ of k is increasing and the function $g(k) - f(g(k))$ of k is also increasing. For clarity, we define a total recursive function $m : \mathbb{N} \to \mathbb{N}$ by $m(k) = f(g(k)) + c$. Since m is injective, it is then easy to see that there exists a partial recursive function $\Phi : \mathbb{N} \to \mathbb{N}$ such that $\Phi(m(k)) = g(k)$ for all $k \in \mathbb{N}$. Therefore, based on the optimality of U, it is shown that

$$K(g(k), \Omega_V\restriction_{m(k)}) \leq K(\Omega_V\restriction_{m(k)}) + O(1)$$

for all $k \in \mathbb{N}$. It follows from (8.7) that $g(k) < K(\Omega_V\restriction_{m(k)}) + O(1)$ for all $k \in \mathbb{N}$. Therefore, using (2.5) we have that $g(k) - f(g(k)) < K(m(k)) + O(1)$ for all $k \in \mathbb{N}$. Then, since the function $g(k) - f(g(k))$ of k is unbounded, it is easy to see that there exists a total recursive function $\Theta : \mathbb{N}^+ \to \mathbb{N}$ such that $l \leq K(\Theta(l))$ for every $l \in \mathbb{N}^+$. It follows from (2.8) that $l \leq K(l) + O(1)$ for all $l \in \mathbb{N}^+$. Hence, using (2.6) we have that $l \leq 2\log_2 l + O(1)$ for all $l \in \mathbb{N}^+$. However, we have a contradiction on letting $l \to \infty$ in this inequality. This completes the proof. $\qquad\square$

8.5 T-Convergent R.E. Reals and Strict T-Compressibility

Let T be an arbitrary computable real with $0 < T \leq 1$. The real parameter T plays a key role throughout the rest of this chapter, as well as in the preceding chapters. In this section, we investigate the relation of T-convergent r.e. reals to the halting problems and to the notion of *strict T-compressibility*, introduced below. In particular, Theorem 8.18 below is used to show Theorem 8.19 in the next section, and plays a major role in establishing the bidirectionality in the next section. On the other hand, Theorem 8.15 below is used to show Theorem 8.20 in the next section.

In Theorem 4.2 we have shown that $Z(T)$ and $F(T)$ are T-compressible in the case where $0 < T \leq 1$ and T is computable. In this section we show that the T-compressibility of $Z(T)$ and $F(T)$ can be strengthened to the *strict T-compressibility* in the case where $0 < T < 1$ and T is computable. Here the notion of strict T-compressibility is defined as follows.

Definition 8.5 *(Strict T-Compressibility, Tadaki [51])* Let T be a real with $0 \leq T \leq 1$. For any $\omega \in \{0, 1\}^\infty$, we say that ω is *strictly T-compressible* if there exists $c \in \mathbb{N}$ such that, for all $n \in \mathbb{N}^+$, it holds that $K(\omega\!\restriction_n) \leq Tn + c$. For any $\alpha \in \mathbb{R}$, we say that α is *strictly T-compressible* if Binary (α) is strictly T-compressible. $\qquad\square$

In 2011, Calude et al. [7] showed the existence of an r.e. real which is weakly Chaitin T-random and strictly T-compressible, in the case where T is a computable real with $0 < T < 1$, as follows.

Theorem 8.12 (Calude et al. [7]) *Suppose that T is a computable real with $0 < T < 1$. Then there exist an r.e. real $\alpha \in (0, 1)$ and $d \in \mathbb{N}$ such that, for all $n \in \mathbb{N}^+$, it holds that $|K(\alpha\!\restriction_n) - Tn| \leq d$.* $\qquad\square$

We prove Theorem 8.12 in a different manner from Calude et al. [7] later. We first show that the same r.e. real α as constructed in the original proof of Theorem 8.12 by Calude et al. [7] has the following property.

Theorem 8.13 *Suppose that T is a computable real with $0 < T < 1$. Let V be an optimal prefix-free machine. Then there exists an r.e. real $\alpha \in (0, 1)$ such that α is weakly Chaitin T-random and $\mathrm{Pf}(\alpha)$ is reducible to $\mathrm{dom}\, V$ in query size $\lfloor Tn \rfloor + O(1)$.* $\qquad\square$

Calude et al. [7] use Lemma 8.2 below to show Theorem 8.12. We also use it to show Theorem 8.13.

Lemma 8.2 (Reimann and Stephan [29] and Calude et al. [7]) *Let T be a real with $T > 0$, and let V be an optimal prefix-free machine.*

(i) *Suppose that $T < 1$. Then there exists $c \in \mathbb{N}^+$ such that, for every $s \in \{0, 1\}^*$, there exists $t \in \{0, 1\}^c$ for which $K_V(st) \geq K_V(s) + Tc$.*

(ii) *There exists $c \in \mathbb{N}^+$ such that, for every $s \in \{0, 1\}^*$, $K_V(s0^c) \leq K_V(s) + Tc - 1$ and $K_V(s1^c) \leq K_V(s) + Tc - 1$.*

Proof Let T be a real with $T > 0$, and let V be an optimal prefix-free machine.
  (i) Chaitin [9] showed that

$$K(s, t) = K(s) + H(t/s) + O(1) \tag{8.8}$$

for all $s, t \in \{0, 1\}^*$. This is Theorem 3.9 (a) of Chaitin [9]. For the definition of $H(t/s)$, see Sect. 2 of Chaitin [9]. We here only use the property that, for every $s \in \{0, 1\}^*$ and every $n \in \mathbb{N}$, there exists $t \in \{0, 1\}^n$ such that

$$H(t/s) \geq n. \tag{8.9}$$

This is easily shown from the definition of $H(t/s)$ by counting the number of binary strings of length less than n.

On the other hand, from the definition of $K(s, t)$ provided in Sect. 2.3, it is easy to show that

$$K(st, |t|) = K(s, t) + O(1) \tag{8.10}$$

for all $s, t \in \{0, 1\}^*$. Then, using (8.1) note that $K(st) + K(|t|) \geq K(st, |t|) - O(1)$ for all $s, t \in \{0, 1\}^*$. Thus, it follows from (8.10), (8.8), and (8.9) that there exists $d \in \mathbb{N}$ such that, for every $s \in \{0, 1\}^*$ and every $n \in \mathbb{N}$, there exists $t \in \{0, 1\}^n$ for which $K(st) \geq K(s) + n - K(n) - d$. Using the optimality of U and V, we then see that there exists $d' \in \mathbb{N}$ such that, for every $s \in \{0, 1\}^*$ and every $n \in \mathbb{N}$, there exists $t \in \{0, 1\}^n$ for which

$$K_V(st) \geq K_V(s) + n - K(n) - d'. \tag{8.11}$$

Now, suppose that $T < 1$. Then, it follows from (2.6) that there exists $c \in \mathbb{N}^+$ such that $K(c) + d' \leq (1 - T)c$. Hence, using (8.11) we have that there exists $c \in \mathbb{N}^+$ such that, for every $s \in \{0, 1\}^*$, there exists $t \in \{0, 1\}^c$ for which $K_V(st) \geq K_V(s) + Tc$.

  (ii) Since V is optimal, it is easy to show that there exists $d \in \mathbb{N}$ such that, for every $s \in \{0, 1\}^*$ and every $n \in \mathbb{N}$,

$$K_V(s0^n) \leq K_V(s) + K(n) + d \quad \text{and} \quad K_V(s1^n) \leq K_V(s) + K(n) + d. \tag{8.12}$$

Since $T > 0$, it follows from (2.6) that there exists $c \in \mathbb{N}^+$ such that $K(c) + d \leq Tc - 1$. Hence, using (8.12) we have that there exists $c \in \mathbb{N}^+$ such that, for every $s \in \{0, 1\}^*$, $K_V(s0^c) \leq K_V(s) + Tc - 1$ and $K_V(s1^c) \leq K_V(s) + Tc - 1$. □

Proof (of Theorem 8.13) Suppose that T is a computable real with $0 < T < 1$. Let V be an optimal prefix-free machine. Then it follows from Lemma 8.2 (i) that there exists $c \in \mathbb{N}^+$ such that, for every $s \in \{0, 1\}^*$, there exists $t \in \{0, 1\}^c$ for which

$$K_V(st) \geq K_V(s) + Tc. \tag{8.13}$$

For each prefix-free machine G and each $s \in \{0, 1\}^*$, we denote by $S(G; s)$ the set $\{ u \in \{0, 1\}^{|s|+c} \mid s \text{ is a prefix of } u \,\&\, K_G(u) > T |u| \}$.

Now, we define a sequence $\{a_k\}_{k \in \mathbb{N}}$ of finite binary strings recursively on $k \in \mathbb{N}$ by $a_k := \lambda$ if $k = 0$ and $a_k := \min S(V; a_{k-1})$ otherwise, where the min is taken with respect to the ordering on $\{0, 1\}^*$ defined in Sect. 2.1. First note that a_0 is properly defined as λ and therefore satisfies $K_V(a_0) > T |a_0|$. For each $k \geq 1$, assume that $a_0, a_1, a_2, \ldots, a_{k-1}$ are properly defined. Then $K_V(a_{k-1}) > T |a_{k-1}|$ holds. It follows from (8.13) that there exists $t \in \{0, 1\}^c$ for which $K_V(a_{k-1}t) \geq K_V(a_{k-1}) + Tc$, and therefore $a_{k-1}t \in \{0, 1\}^{|a_{k-1}|+c}$ and $K_V(a_{k-1}t) > T |a_{k-1}t|$. Thus $S(V; a_{k-1}) \neq \emptyset$, and therefore a_k is properly defined. Hence, a_k is properly defined for every $k \in \mathbb{N}$. We thus see that, for every $k \in \mathbb{N}$, it holds that $a_k \in \{0, 1\}^{ck}$, $K_V(a_k) > T |a_k|$, and a_k is a prefix of a_{k+1}.

Thus, we can uniquely define an infinite binary sequence ω by the condition that $\omega{\restriction}ck = a_k$ for all $k \in \mathbb{N}^+$. It follows that $K_V(\omega{\restriction}ck) > T |\omega{\restriction}ck|$ for all $k \in \mathbb{N}^+$. Note that, for every $t \in \{0, 1\}^*$, there exists $d_t \in \mathbb{N}$ such that, for every $s \in \{0, 1\}^*$, it holds that $|K_V(st) - K_V(s)| \leq d_t$. Therefore there exists $d \in \mathbb{N}$ such that, for every $s, t \in \{0, 1\}^*$, if $|t| \leq c$ then $|K_V(st) - K_V(s)| \leq d$. Thus, we have that $K_V(\omega{\restriction}n) > Tn - d$ for all $n \in \mathbb{N}^+$, which implies that ω is weakly Chaitin T-random.

We define a real α by $\alpha := 0.\omega$. Since ω is weakly Chaitin T-random, the real α is irrational and its base-two expansion ω contains infinitely many zeros. Therefore $\alpha \in (0, 1)$ and Binary $(\alpha) = \omega$. Thus, α is weakly Chaitin T-random.

Next, we show that $\mathrm{Pf}(\alpha)$ is reducible to dom V in query size $\lfloor Tn \rfloor + O(1)$. For each $k \in \mathbb{N}$, we denote by F_k the set $\{s \in \{0, 1\}^* \mid K_V(s) \leq \lfloor Tck \rfloor\}$. It follows that

$$a_k = \min\{ u \in \{0, 1\}^{ck} \mid a_{k-1} \text{ is a prefix of } u \,\&\, u \notin F_k \} \tag{8.14}$$

for every $k \in \mathbb{N}^+$. Then, by the following procedure, we see that $\mathrm{Pf}(\alpha)$ is reducible to dom V in query size $\lfloor Tn \rfloor + O(1)$.

Given $s \in \{0, 1\}^*$ with $s \neq \lambda$, one first calculates the k_0 finite sets $F_1, F_2, \ldots, F_{k_0}$, where $k_0 = \lceil |s|/c \rceil$, by putting queries to the oracle dom V. Note here that all the queries can be of length at most $\lfloor T(|s| + c) \rfloor$. One then calculates $a_1, a_2, \ldots, a_{k_0}$ one by one in this order from $a_0 = \lambda$, based on the relation (8.14) and $F_1, F_2, \ldots, F_{k_0}$. Finally, one accepts s if s is a prefix of a_{k_0} and rejects otherwise. This is possible since $\alpha{\restriction}ck_0 = a_{k_0}$ and $|s| \leq ck_0$.

Finally, we show that the real α is r.e. Let $p_1, p_2, p_3, \ldots$ be a particular recursive enumeration of the infinite r.e. set dom V. For each $l \in \mathbb{N}^+$, we define a prefix-free machine $V^{(l)}$ by the conditions that (i) dom $V^{(l)} = \{p_1, p_2, \ldots, p_l\}$, and (ii) $V^{(l)}(p) = V(p)$ for every $p \in \text{dom } V^{(l)}$. It is easy to see that such prefix-free machines $V^{(1)}, V^{(2)}, V^{(3)}, \ldots$ exist. For each $l \in \mathbb{N}^+$ and $s \in \{0, 1\}^*$, note that

$$K_{V^{(l)}}(s) \geq K_V(s) \tag{8.15}$$

holds, where $K_{V^{(l)}}(s)$ may be ∞. For each $l \in \mathbb{N}^+$, we define a sequence $\{a_k^{(l)}\}_{k \in \mathbb{N}}$ of finite binary strings recursively on $k \in \mathbb{N}$ by $a_k^{(l)} := \lambda$ if $k = 0$ and $a_k^{(l)} :=$ $\min(S(V^{(l)}; a_{k-1}^{(l)}) \cup \{a_{k-1}^{(l)} 1^c\})$ otherwise. It follows that $a_k^{(l)}$ is properly defined for every $l \in \mathbb{N}^+$ and $k \in \mathbb{N}$. Note, in particular, that $a_k^{(l)} \in \{0, 1\}^{ck}$ and $a_k^{(l)}$ is a prefix of $a_{k+1}^{(l)}$ for every $l \in \mathbb{N}^+$ and $k \in \mathbb{N}$.

Let $l \in \mathbb{N}^+$. We show that $a_k^{(l)} \leq a_k$ for every $k \in \mathbb{N}$, where $a_k^{(l)}$ and a_k are regarded as a dyadic integer. In the case where $a_k^{(l)} = a_k$ for every $k \in \mathbb{N}$, the result is obvious. Thus, we assume that $a_k^{(l)} \neq a_k$ for some $k \in \mathbb{N}$. Then, let m be the smallest $k \in \mathbb{N}$ such that $a_k^{(l)} \neq a_k$. Since $a_0^{(l)} = \lambda = a_0$, it follows that $m \geq 1$ and $a_{m-1}^{(l)} = a_{m-1}$. Then, based on the constructions of $a_m^{(l)}$ and a_m from $a_{m-1}^{(l)}$ and a_{m-1}, respectively, and the inequality (8.15), we see that $a_m^{(l)} \leq a_m$. However, since $a_m^{(l)} \neq a_m$ from the definition of m, we have $a_m^{(l)} < a_m$. Therefore, for each $k > m$, since $a_m^{(l)}$ is a prefix of $a_k^{(l)}$ and a_m is a prefix of a_k, we have $a_k^{(l)} < a_k$. Thus, we see that $a_k^{(l)} \leq a_k$ for every $k \in \mathbb{N}$, as desired.

We define a sequence $\{r_k\}_{k \in \mathbb{N}}$ of rationals by $r_k := 0.a_k^{(k)}$. Obviously, $\{r_k\}_{k \in \mathbb{N}}$ is a computable sequence of rationals. Note that $\alpha\lceil_{ck} = \omega\lceil_{ck} = a_k$ for all $k \in \mathbb{N}^+$, and $\alpha \in (0, 1)$. Thus, based on the result in the previous paragraph, we see that $r_k \leq \alpha$ for all $k \in \mathbb{N}$. Moreover, based on the constructions of the prefix-free machines $V^{(1)}, V^{(2)}, V^{(3)}, \ldots$ from V, it is easy to see that $\lim_{k \to \infty} r_k = \alpha$. Hence, the real α is r.e. $\qquad\square$

Using Theorems 8.3 and 8.13, we can give to Theorem 8.12 a different proof from the original one given in Calude et al. [7] as follows.

Proof (of Theorem 8.12 in a different manner from Calude et al. [7]) Suppose that T is a computable real with $0 < T < 1$. We choose a particular optimal prefix-free machine V and a particular deterministic Turing machine M which computes V. For each $l \geq L_M$, we choose a particular p_l from I_M^l. By Theorem 8.13, there exist an r.e. real $\alpha \in (0, 1)$ and $c \in \mathbb{N}^+$ such that α is weakly Chaitin T-random and $\mathrm{Pf}(\alpha)$ is reducible to dom V in query size $\lfloor Tn \rfloor + c$. In what follows, we show that α is strictly T-compressible.

We first define a total recursive function $f : \mathbb{N} \to \mathbb{N}$ by $f(n) = \lfloor Tn \rfloor + c$. Then, given $(n, p_{f(n)})$ with $n \in \mathbb{N}^+$ and $f(n) \geq L_M$, one can calculate the finite set dom $V\lceil_{f(n)}$ by simulating the computation of M with the input q until at most the time step $T_M(p_{f(n)})$, for each $q \in \{0, 1\}^*$ with $|q| \leq f(n)$. This can be possible because $T_M(p_{f(n)}) = T_M^{f(n)}$ for every $n \in \mathbb{N}^+$ with $f(n) \geq L_M$. Thus, since $\mathrm{Pf}(\alpha)$ is reducible to dom V in query size f, we see that there exists a partial recursive function $\Psi : \mathbb{N} \times \{0, 1\}^* \to \{0, 1\}^*$ such that $\Psi(n, p_{f(n)}) = \alpha\lceil_n$ for every $n \in \mathbb{N}^+$ with $f(n) \geq L_M$. It follows from the inequality (2.8) that

$$K(\alpha\lceil_n) \leq K(n, p_{f(n)}) + O(1) \tag{8.16}$$

for all $n \in \mathbb{N}^+$ with $f(n) \geq L_M$.

On the other hand, given $f(n)$ with $n \in \mathbb{N}^+$, one only needs to specify one of $\lceil 1/T \rceil$ possibilities of n in order to calculate n. This is possible since T is a computable real with $T > 0$ and $\lceil (f(n) - c)/T \rceil \leq n \leq \lceil (f(n) - c)/T \rceil + \lceil 1/T \rceil - 1$ holds. Thus, there exists a partial recursive function $\Phi \colon \mathbb{N}^+ \times \{0, 1\}^* \times \mathbb{N}^+ \to \mathbb{N}^+ \times \{0, 1\}^*$ such that, for every $n \in \mathbb{N}^+$ and $p \in \{0, 1\}^*$, there exists $k \in \mathbb{N}^+$ with the properties that $1 \leq k \leq \lceil 1/T \rceil$ and

$$\Phi(f(n), p, k) = (n, p). \tag{8.17}$$

Then, we can show that there exists a partial recursive function $\Psi \colon \{0, 1\}^* \to \{0, 1\}^*$ such that, for every $n \in \mathbb{N}^+$, $p \in \{0, 1\}^*$, and $k \in \mathbb{N}^+$ with $1 \leq k \leq \lceil 1/T \rceil$, it holds that $\Psi(b(b(f(n), p), k)) = \Phi(f(n), p, k)$, where b is the 'standard' bijective total recursive function from $\{0, 1\}^* \times \{0, 1\}^*$ to $\{0, 1\}^*$ introduced in Sect. 2.3. Applying (2.8) to this Ψ, we have that $K(\Phi(f(n), p, k)) \leq K(b(b(f(n), p), k) + O(1)$ for all $n \in \mathbb{N}^+$, $p \in \{0, 1\}^*$, and $k \in \mathbb{N}^+$ with $1 \leq k \leq \lceil 1/T \rceil$. Thus, it follows from (8.17) and (8.1) that

$$K(n, p) \leq K(f(n), p) + \max\{K(k) \mid k \in \mathbb{N}^+ \ \& \ 1 \leq k \leq \lceil 1/T \rceil\} + O(1)$$

for all $n \in \mathbb{N}^+$ and $p \in \{0, 1\}^*$. Hence, using (8.16) and Theorem 8.3 we have that

$$K(\alpha\restriction_n) \leq K(f(n), p_{f(n)}) + O(1) \leq f(n) + O(1) \leq Tn + O(1)$$

for all $n \in \mathbb{N}^+$ with $f(n) \geq L_M$. It follows that $K(\alpha\restriction_n) \leq Tn + O(1)$ for all $n \in \mathbb{N}^+$, which implies that α is strictly T-compressible. This completes the proof. $\qquad\square$

Using Theorems 8.12 and 7.5 we can prove Theorem 8.14 below, which strengthens Theorem 7.12 in the case of $T < 1$.

Theorem 8.14 *Suppose that T is a computable real with $0 < T < 1$. For every r.e. real β, if β is T-convergent then β is strictly T-compressible.*

Proof Suppose that T is a computable real with $0 < T < 1$. It follows from Theorem 8.12 that there exist an r.e. real $\alpha \in (0, 1)$ and $d_0 \in \mathbb{N}$ such that α is weakly Chaitin T-random and

$$K(\alpha\restriction_n) \leq Tn + d_0 \tag{8.18}$$

for all $n \in \mathbb{N}^+$. Since α is weakly Chaitin T-random and r.e., using the implication (i) $\Rightarrow$ (vi) of Theorem 7.5 we have that, for every T-convergent r.e. real β, there exists $d \in \mathbb{N}$ such that, for all $n \in \mathbb{N}^+$, it holds that $K(\beta\restriction_n) \leq K(\alpha\restriction_n) + d$. Thus, for each T-convergent r.e. real β, using (8.18) we see that $K(\beta\restriction_n) \leq Tn + O(1)$ for all $n \in \mathbb{N}^+$, which implies that β is strictly T-compressible. $\qquad\square$

Now, as one of the main applications of Theorem 8.14, we can prove the following theorem for the partition function $Z(T)$ and the Helmholtz free energy $F(T)$ of AIT.

Theorem 8.15 *Suppose that T is a computable real with* $0 < T < 1$. *Let V be an optimal prefix-free machine. Then, each of* $Z_V(T)$ *and* $F_V(T)$ *is weakly Chaitin T-random and strictly T-compressible. In other words,* $K(Z_V(T)\lceil_n) = Tn + O(1)$ *and* $K(F_V(T)\lceil_n) = Tn + O(1)$ *hold for all* $n \in \mathbb{N}^+$.

Proof Suppose that T is a computable real with $0 < T < 1$. Let V be an optimal prefix-free machine. By Theorem 5.31 (i), each of $Z_V(T)$ and $F_V(T)$ is weakly Chaitin T-random. On the other hand, since each of $Z_V(T)$ and $-F_V(T)$ is a T-convergent r.e. real due to Theorem 7.4, it follows from Theorem 8.14 that each of $Z_V(T)$ and $-F_V(T)$ is strictly T-compressible. Thus, since

$$K((-F_V(T))\lceil_n) = K(F_V(T)\lceil_n) + O(1)$$

holds for all $n \in \mathbb{N}^+$, we have that $F_V(T)$ is strictly T-compressible.   □

On the other hand, Theorem 8.14 results also in the following, obviously.

Corollary 8.1 *Suppose that T is a computable real with* $0 < T < 1$. *For every r.e. real* β, *if* β *is Chaitin T-random then* β *is not T-convergent.*   □

Using Corollary 8.1, we can show the following theorem for the energy $F(T)$, the entropy $S(T)$, and the specific heat $C(T)$ of AIT, as well as for $W(Q, T)$.

Theorem 8.16 *Suppose that T is a computable real with* $0 < T < 1$. *Let V be an optimal prefix-free machine, and let Q be a computable real with* $Q > 0$. *Then, neither* $E_V(T)$, $S_V(T)$, $C_V(T)$ *nor* $W(Q, T)$ *is T-convergent.*

Proof By Theorem 7.1 (i), Theorem 5.32 (i), and Theorem 3.4 (i), we see that all of $E_V(T)$, $S_V(T)$, $C_V(T)$, and $W(Q, T)$ are r.e. and Chaitin T-random. Thus, the result follows immediately from Corollary 8.1.   □

Calude et al. [7], in essence, showed the following result. For completeness, we include its proof.

Theorem 8.17 (Calude et al. [7]) *If a real* β *is weakly Chaitin T-random and strictly T-compressible, then there exists* $d \geq 3$ *such that the base-two expansion of* β, *i.e.,* Binary (β), *has neither a run of d consecutive zeros nor a run of d consecutive ones.*

Proof Let β be a real which is weakly Chaitin T-random and strictly T-compressible. Then there exists $d_0 \in \mathbb{N}^+$ such that, for every $n \in \mathbb{N}^+$, it holds that

$$|K(\beta\lceil_n) - Tn| \leq d_0. \tag{8.19}$$

On the other hand, by Lemma 8.2 (ii) we see that there exists $c \in \mathbb{N}^+$ such that, for every $s \in \{0, 1\}^*$, it holds that

$$K(s0^c) \leq K(s) + Tc - 1 \tag{8.20}$$

and

$$K(s1^c) \le K(s) + Tc - 1. \tag{8.21}$$

We choose a particular $k_0 \in \mathbb{N}^+$ with $k_0 > 2d_0$, and set $d := ck_0$. Note that $d \ge 3$.

Let us assume contrarily that Binary (β) has a run of d consecutive zeros. Then $\beta\lceil_{n_0} 0^d = \beta\lceil_{n_0+d}$ for some $n_0 \in \mathbb{N}$. Thus, using (8.20) repeatedly, we have that

$$K(\beta\lceil_{n_0+d}) - T(n_0 + d) + k_0 \le K(\beta\lceil_{n_0}) - Tn_0.$$

Therefore, using the triangle inequality, we have that

$$k_0 \le \left| K(\beta\lceil_{n_0}) - Tn_0 \right| + \left| K(\beta\lceil_{n_0+d}) - T(n_0 + d) \right|.$$

It follows from (8.19) that $k_0 \le 2d_0$. However, this contradicts the choice of k_0 such that $k_0 > 2d_0$. Hence, Binary (β) does not have a run of d consecutive zeros.

In a similar manner, using (8.21) we can show that Binary (β) does not have a run of d consecutive ones, as well. This completes the proof. □

The following theorem plays a key role in the next section.

Theorem 8.18 *Suppose that T is a computable real with $0 < T < 1$. Let V be an optimal prefix-free machine. For every r.e. real β, if β is T-convergent and weakly Chaitin T-random, then $\mathrm{Pf}(\beta)$ is reducible to $\mathrm{dom}\, V$ in query size $\lfloor Tn \rfloor + O(1)$.*

Proof Suppose that T is a computable real with $0 < T < 1$. Let V be an optimal prefix-free machine. Then, by Theorem 8.13, there exist an r.e. real $\alpha \in (0, 1)$ and $d_0 \in \mathbb{N}$ such that α is weakly Chaitin T-random and $\mathrm{Pf}(\alpha)$ is reducible to $\mathrm{dom}\, V$ in query size $\lfloor Tn \rfloor + d_0$. Since α is an r.e. real which is weakly Chaitin T-random, it follow from the implication (i) $\Rightarrow$ (iii) of Theorem 7.5 that α is $\Omega(T)$-like.

Now, for an arbitrary r.e. real β, assume that β is T-convergent and weakly Chaitin T-random. Then, by Theorem 8.14, β is strictly T-compressible. It follows from Theorem 8.17 that there exists $c \ge 3$ such that Binary (β) has neither a run of c consecutive zeros nor a run of c consecutive ones. On the other hand, since the r.e. real α is $\Omega(T)$-like, from the definition of $\Omega(T)$-likeness, i.e., Definition 7.4, we see that α dominates β. Therefore, there are computable, increasing sequences $\{a_k\}_{k \in \mathbb{N}}$ and $\{b_k\}_{k \in \mathbb{N}}$ of rationals and $d_1 \in \mathbb{N}$ such that (i) $\lim_{k \to \infty} a_k = \alpha$, (ii) $\lim_{k \to \infty} b_k = \beta$, and (iii) $\alpha - a_k \ge 2^{-d_1}(\beta - b_k)$ and $\lfloor \beta \rfloor = \lfloor b_k \rfloor$ for all $k \in \mathbb{N}$. We set $d_2 := d_1 + c + 2$. Then, by the following procedure, we see that $\mathrm{Pf}(\beta)$ is reducible to $\mathrm{dom}\, V$ in query size $\lfloor T(n + d_2) \rfloor + d_0$.

Given $s \in \{0, 1\}^*$, one first calculates $\alpha\lceil_{m+d_2}$ by putting to the oracle $\mathrm{dom}\, V$ the queries of length at most $\lfloor T(m + d_2) \rfloor + d_0$, where $m := |s|$. This is possible since $\mathrm{Pf}(\alpha)$ is reducible to $\mathrm{dom}\, V$ in query size $\lfloor Tn \rfloor + d_0$. One then find $k_0 \in \mathbb{N}$ such that $0.(\alpha\lceil_{m+d_2}) < a_{k_0}$. This is possible since $0.(\alpha\lceil_{m+d_2}) < \alpha$ and $\lim_{k \to \infty} a_k = \alpha$. It follows that

$$2^{-(m+d_2)} > \alpha - 0.(\alpha\lceil_{m+d_2}) > \alpha - a_{k_0} \ge 2^{-d_1}(\beta - b_{k_0}).$$

Thus, $0 < \beta - b_{k_0} < 2^{-(m+c+2)}$. Note that $|\beta - \lfloor\beta\rfloor - 0.(\beta\lceil_{m+c+2})| < 2^{-(m+c+2)}$ and $|b_{k_0} - \lfloor b_{k_0}\rfloor - 0.t| \leq 2^{-(m+c+2)}$, where t denotes $b_{k_0}\lceil_{m+c+2}$. Since $\lfloor\beta\rfloor = \lfloor b_{k_0}\rfloor$ it follows that $|0.(\beta\lceil_{m+c+2}) - 0.t| < 3 \cdot 2^{-(m+c+2)}$. Hence, $|\beta\lceil_{m+c+2} - t| \leq 2$, where $\beta\lceil_{m+c+2}$ and t in $\{0, 1\}^{m+c+2}$ are regarded as a dyadic integer. Thus, t is obtained by adding to $\beta\lceil_{m+c+2}$ or subtracting from $\beta\lceil_{m+c+2}$ a 2-bit integer, which is either 00, 01, or 10. Since Binary (β) has neither a run of c consecutive zeros nor a run of c consecutive ones, note that every consecutive c bits occurring in Binary (β) contains both zero and one. It follows that the first m bits of t equals to $\beta\lceil_m$. Thus, one accepts s if s is a prefix of t and rejects otherwise. Recall here that $|s| = m$. □

8.6 Bidirectionality

In this section we show the bidirectionality between $Z_U(T)$ and dom U with a computable real $T \in (0, 1)$ in a general setting. Theorems 8.19 and 8.20 below are two of the main results of this chapter.

Theorem 8.19 (Elaboration of $Z_U(T) \leq_{wtt}$ dom U) *Suppose that T is a computable real with $0 < T < 1$. Let V and W be optimal prefix-free machines, and let f be an order function. Then the following two conditions are equivalent:*

(i) $\mathrm{Pf}(Z_V(T))$ *is reducible to* dom W *in query size* $f(n) + O(1)$.
(ii) $Tn \leq f(n) + O(1)$. □

Theorem 8.20 (Elaboration of dom $U \leq_{wtt} Z_U(T)$) *Suppose that T is a computable real with $0 < T \leq 1$. Let V and W be optimal prefix-free machines, and let f be an order function. Then the following two conditions are equivalent:*

(i) dom W *is reducible to* $\mathrm{Pf}(Z_V(T))$ *in query size* $f(n) + O(1)$.
(ii) $n/T \leq f(n) + O(1)$. □

Theorem 8.19 and Theorem 8.20 are proved in Sect. 8.6.1 and Sect. 8.6.2 below, respectively. Note that the function Tn in the condition (ii) of Theorem 8.19 and the function n/T in the condition (ii) of Theorem 8.20 are the inverse functions of each other. This implies that the computations between $\mathrm{Pf}(Z_V(T))$ and dom W are bidirectional in the case where T is a computable real with $0 < T < 1$. The formal proof is as follows.

Theorem 8.21 *Suppose that T is a computable real with $0 < T < 1$. Let V and W be optimal prefix-free machines. Then the computations between $\mathrm{Pf}(Z_V(T))$ and dom W are bidirectional.*

Proof Let V and W be optimal prefix-free machines. It follows from the implication (ii) $\Rightarrow$ (i) of Theorem 8.20 that there exists $c \in \mathbb{N}$ for which dom W is reducible to $\mathrm{Pf}(Z_V(T))$ in query size f with $f(n) = \lfloor n/T \rfloor + c$. On the other hand, it follows from the implication (ii) $\Rightarrow$ (i) of Theorem 8.19 that there exists $d \in \mathbb{N}$ for which

$\mathrm{Pf}(Z_V(T))$ is reducible to dom W in query size g with $g(n) = \lfloor Tn \rfloor + d$. Since T is computable, f and g are order functions. For each $n \in \mathbb{N}$, we see that $g(f(n)) \leq Tf(n) + d \leq n + Tc + d$. Thus, the computation from $\mathrm{Pf}(Z_V(T))$ to dom W is not unidirectional. In a similar manner, we see that the computation from dom W to $\mathrm{Pf}(Z_V(T))$ is not unidirectional. This completes the proof. $\qquad\square$

8.6.1   The Proof of Theorem 8.19

Let T be a computable real with $0 < T < 1$, and let V be an optimal prefix-free machine. Then, by Theorem 7.4, $Z_V(T)$ is a T-convergent r.e. real. Moreover, by Theorem 5.31 (i), $Z_V(T)$ is weakly Chaitin T-random. Thus, the implication (ii) $\Rightarrow$ (i) of Theorem 8.19 follows immediately from Theorem 8.18 and Proposition 8.1 (ii).

On the other hand, the implication (i) $\Rightarrow$ (ii) of Theorem 8.19 follows immediately from Theorem 8.22 below and Theorem 5.31 (i). In order to prove Theorem 8.22, we use Theorem 8.3.

Theorem 8.22 *Suppose that T is a computable real with $0 < T \leq 1$. Let β be a real which is weakly Chaitin T-random, and let V be an optimal prefix-free machine. For every order function f, if $\mathrm{Pf}(\beta)$ is reducible to dom V in query size f then $Tn \leq f(n) + O(1)$.*

Proof Suppose that T is a computable real with $0 < T \leq 1$. Let β be a real which is weakly Chaitin T-random, and let V be an optimal prefix-free machine. For an arbitrary order function f, assume that $\mathrm{Pf}(\beta)$ is reducible to dom V in query size f. We choose a particular deterministic Turing machine M which computes V. For each $l \geq L_M$, we then choose a particular p_l from I_M^l. Then, by the following procedure, we see that there exists a partial recursive function $\Psi \colon \mathbb{N} \times \{0, 1\}^* \to \{0, 1\}^*$ such that, for every $n \in \mathbb{N}^+$ with $f(n) \geq L_M$,

$$\Psi(n, p_{f(n)}) = \beta\!\restriction_n . \tag{8.22}$$

Given $(n, p_{f(n)})$ with $n \in \mathbb{N}^+$ and $f(n) \geq L_M$, one first calculates the finite set dom $V\!\restriction_{f(n)}$ by simulating the computation of M with the input q until at most the time step $T_M(p_{f(n)})$, for each $q \in \{0, 1\}^*$ with $|q| \leq f(n)$. This can be possible because $T_M(p_{f(n)}) = T_M^{f(n)}$ for every $n \in \mathbb{N}^+$ with $f(n) \geq L_M$. One then calculates $\beta\!\restriction_n$ using dom $V\!\restriction_{f(n)}$ and outputs it. This is possible since $\mathrm{Pf}(\beta)$ is reducible to dom V in query size f.

It follows from (8.22) and (2.8) that

$$K(\beta\!\restriction_n) \leq K(n, p_{f(n)}) + O(1) \tag{8.23}$$

for all $n \in \mathbb{N}^+$ with $f(n) \geq L_M$.

Now, let us assume contrarily that the function $Tn - f(n)$ of $n \in \mathbb{N}$ is unbounded. Recall that f is an order function and T is computable. Hence, it is easy to show that there exists a total recursive function $g: \mathbb{N} \to \mathbb{N}$ such that the function $f(g(k))$ of k is increasing and the function $Tg(k) - f(g(k))$ of k is also increasing. Since the function $f(g(k))$ of k is injective, it is then easy to see that there exists a partial recursive function $\Phi: \mathbb{N} \to \mathbb{N}$ such that $\Phi(f(g(k))) = g(k)$ for all $k \in \mathbb{N}$. Thus, it follows from the optimality of U that $K(g(k), s) \le K(f(g(k)), s) + O(1)$ for all $k \in \mathbb{N}$ and $s \in \{0, 1\}^*$. Hence, using (8.23) and Theorem 8.3 we have that

$$K(\beta\!\restriction_{g(k)}) \le K(g(k), p_{f(g(k))}) + O(1) \le K(f(g(k)), p_{f(g(k))}) + O(1)$$
$$\le f(g(k)) + O(1)$$

for all $k \in \mathbb{N}^+$ with $f(g(k)) \ge L_M$. Then, since β is weakly Chaitin T-random, we have that

$$Tg(k) \le K(\beta\!\restriction_{g(k)}) + O(1) \le f(g(k)) + O(1)$$

for all $k \in \mathbb{N}^+$ with $f(g(k)) \ge L_M$. However, this contradicts the property that the function $Tg(k) - f(g(k))$ of k is unbounded, and the proof is completed. $\square$

8.6.2 The Proof of Theorem 8.20

The implication (i) $\Rightarrow$ (ii) of Theorem 8.20 can be proved based on Theorem 8.15 as follows.

Proof (of (i) $\Rightarrow$ (ii) of Theorem 8.20) In the case of $T = 1$, the implication (i) $\Rightarrow$ (ii) of Theorem 8.20 results in the implication (i) $\Rightarrow$ (ii) of Theorem 8.5. Thus, we assume that T is a computable real with $0 < T < 1$ in what follows. Let V and W be optimal prefix-free machines, and let f is an order function. Suppose that there exists $c \in \mathbb{N}$ such that dom W is reducible to $\mathrm{Pf}(Z_V(T))$ in query size $f(n) + c$. Then, by considering the following procedure, we first see that $n < K(n, Z_V(T)\!\restriction_{f(n)+c}) + O(1)$ for all $n \in \mathbb{N}$.

Given n and $Z_V(T)\!\restriction_{f(n)+c}$, one first calculates the finite set dom $W\!\restriction_n$. This is possible since dom W is reducible to $\mathrm{Pf}(Z_V(T))$ in query size $f(n) + c$ and $f(k) \le f(n)$ for all $k \le n$. Then, by calculating the set $\{ W(p) \mid p \in \mathrm{dom}\, W\!\restriction_n\}$ and picking any particular finite binary string which is not in this set, one can obtain an $s \in \{0, 1\}^*$ such that $n < K_W(s)$.

Thus, there exists a partial recursive function $\Psi: \mathbb{N} \times \{0, 1\}^* \to \{0, 1\}^*$ such that $n < K_W(\Psi(n, Z_V(T)\!\restriction_{f(n)+c}))$ for every $n \in \mathbb{N}$. It follows from the optimality of W that

$$n < K(n, Z_V(T)\!\restriction_{f(n)+c}) + O(1) \tag{8.24}$$

for all $n \in \mathbb{N}$.

Now, let us assume contrarily that the function $n/T - f(n)$ of $n \in \mathbb{N}$ is unbounded. Recall that f is an order function and T is computable. Hence, it is easy to show that there exists a total recursive function $g: \mathbb{N} \to \mathbb{N}$ such that the function $f(g(k))$ of k

is increasing and the function $g(k)/T - f(g(k))$ of k is also increasing. For clarity, we define a total recursive function $m: \mathbb{N} \to \mathbb{N}$ by $m(k) = f(g(k)) + c$. Since m is injective, it is then easy to see that there exists a partial recursive function $\Phi: \mathbb{N} \to \mathbb{N}$ such that $\Phi(m(k)) = g(k)$ for all $k \in \mathbb{N}$. Therefore, based on the optimality of U, it is shown that

$$K(g(k), Z_V(T){\upharpoonright}_{m(k)}) \leq K(Z_V(T){\upharpoonright}_{m(k)}) + O(1)$$

for all $k \in \mathbb{N}$. It follows from (8.24) that $g(k) < K(Z_V(T){\upharpoonright}_{m(k)}) + O(1)$ for all $k \in \mathbb{N}$. On the other hand, since T is a computable real with $0 < T < 1$, it follows from Theorem 8.15 that $K(Z_V(T){\upharpoonright}_n) \leq Tn + O(1)$ for all $n \in \mathbb{N}$. Therefore we have $g(k) < Tf(g(k)) + O(1)$ for all $k \in \mathbb{N}$. However, this contradicts the property that the function $g(k)/T - f(g(k))$ of k is unbounded, and the proof is completed. $\square$

On the other hand, the implication (ii) $\Rightarrow$ (i) of Theorem 8.20 follows immediately from Theorem 8.23 below and Proposition 8.1 (ii).

Theorem 8.23 *Suppose that T is a computable real with $0 < T \leq 1$. Let V be an optimal prefix-free machine, and let F be a prefix-free machine. Then $\mathrm{dom}\, F$ is reducible to $\mathrm{Pf}(Z_V(T))$ in query size $\lceil n/T \rceil + O(1)$.*

Proof In the case where $\mathrm{dom}\, F$ is a finite set, the result is obvious. Thus, in what follows, we assume that $\mathrm{dom}\, F$ is an infinite set.

Let $p_0, p_1, p_2, p_3, \ldots$ be a particular recursive enumeration of $\mathrm{dom}\, F$, and let G be a prefix-free machine such that $\mathrm{dom}\, G = \mathrm{dom}\, F$ and $G(p_i) = i$ for all $i \in \mathbb{N}$. Recall here that we identify $\{0, 1\}^*$ with $\mathbb{N}$. It is easy to see that such a prefix-free machine G exists. Since V is an optimal prefix-free machine, it follows from the definition of the optimality that there exists $d \in \mathbb{N}$ such that, for every $i \in \mathbb{N}$, there exists $q \in \{0, 1\}^*$ for which $V(q) = G(p_i)$ and $|q| \leq |p_i| + Td$. Thus, $K_V(i) \leq |p_i| + Td$ for every $i \in \mathbb{N}$. For each $s \in \{0, 1\}^*$, we define $Z_V(T; s)$ as $\sum_{V(p)=s} 2^{-|p|/T}$. Then, for each $i \in \mathbb{N}$,

$$Z_V(T; i) \geq 2^{-K_V(i)/T} \geq 2^{-|p_i|/T-d}. \tag{8.25}$$

Then, by the following procedure, we see that $\mathrm{dom}\, F$ is reducible to $\mathrm{Pf}(Z_V(T))$ in query size $\lceil n/T \rceil + d$.

Given $s \in \{0, 1\}^*$, one first calculates $Z_V(T){\upharpoonright}_{\lceil n/T \rceil + d}$ by putting the queries t to the oracle $\mathrm{Pf}(Z_V(T))$ for all $t \in \{0, 1\}^{\lceil n/T \rceil + d}$, where $n = |s|$. Note here that all the queries are of length $\lceil n/T \rceil + d$. One then finds $k_e \in \mathbb{N}$ such that $\sum_{i=0}^{k_e} Z_V(T; i) > 0.(Z_V(T){\upharpoonright}_{\lceil n/T \rceil + d})$. This is possible because $0.(Z_V(T){\upharpoonright}_{\lceil n/T \rceil + d}) < Z_V(T)$, $\lim_{k \to \infty} \sum_{i=0}^{k} Z_V(T; i) = Z_V(T)$, and T is a computable real. It follows that

$$\sum_{i=k_e+1}^{\infty} Z_V(T; i) = Z_V(T) - \sum_{i=0}^{k_e} Z_V(T; i) < Z_V(T) - 0.(Z_V(T){\upharpoonright}_{\lceil n/T \rceil + d}) \leq 2^{-n/T-d}.$$

Therefore, by (8.25),

$$\sum_{i=k_e+1}^{\infty} 2^{-|p_i|/T} \le 2^d \sum_{i=k_e+1}^{\infty} Z_V(T; i) < 2^{-n/T}.$$

It follows that, for every $i > k_e$, $2^{-|p_i|/T} < 2^{-n/T}$ and therefore $n < |p_i|$. Hence, dom $F\!\restriction_n = \{\, p_i \mid i \le k_e \ \& \ |p_i| \le n \,\}$. Thus, one can calculate the finite set dom $F\!\restriction_n$. Finally, one accepts s if $s \in$ dom $F\!\restriction_n$ and rejects otherwise. $\qquad\square$

8.7 Concluding Remarks

Suppose that T is a computable real with $0 < T \le 1$. Let V and W be optimal prefix-free machines. It is worthwhile to clarify the origin of the difference of the behavior of $Z_V(T)$ between $T = 1$ and $T < 1$ with respect to the notion of reducibility in query size f. In the case of $T = 1$, the Ample Excess Lemma [26] (i.e., Theorem 8.10) plays a major role in establishing the unidirectionality of the computation from Ω_V to dom W. However, in the case of $T < 1$, this is not true because the weak Chaitin T-randomness of a real α does not necessarily imply that $\sum_{n=1}^{\infty} 2^{Tn-K(\alpha\restriction_n)} < \infty$, due to Reimann and Stephan [29] (see Sect. 3.1). On the other hand, in the case of $T < 1$, Lemma 8.2 (i) plays a major role in establishing the bidirectionality of the computations between $Z_V(T)$ and dom W. However, this does not hold for the case of $T = 1$.

Chapter 9
Other Related Results and Future Development

9.1 Current Status of the Research

In addition to the works described in the preceding chapters, we did various works about the statistical mechanical interpretation of AIT. This section summarizes some of them. At present, we are trying various approaches to reveal the overall picture of the statistical mechanical interpretation of AIT.

9.1.1 Clarification of the Property of Fixed Points by Statistical Mechanical Technique

Tadaki [53] introduced the notion of the *composition of quantum systems* into AIT, and then clarified a property of the fixed points on partial randomness, called *simultaneous disjointness*, based on the notion of the composition.

Let V be an arbitrary optimal prefix-free machine. In Sect. 5.5 we have defined the sets $\mathscr{Z}(V), \mathscr{F}(V), \mathscr{E}(V)$, and $\mathscr{S}(V)$ based on the computability of the partition function $Z_V(T)$, the Helmholtz free energy $F_V(T)$, the energy $E_V(T)$, and the entropy $S_V(T)$, respectively. Then, we see from Theorem 5.27 that $\mathscr{Z}(V) \cap \mathscr{F}(V) = \emptyset$. By generalizing the technique used in the proof of Theorem 5.27, Tadaki [53] proved the following theorem which states the mutual exclusiveness of these sets.

Theorem 9.1 (Simultaneous Disjointness, Tadaki [53]) *There exists a recursive infinite sequence $V_1, V_2, V_3, \ldots$ of optimal prefix-free machines such that*

$$\mathscr{Z}(V_k) \cap \mathscr{Z}(V_l) = \mathscr{F}(V_k) \cap \mathscr{F}(V_l) = \mathscr{E}(V_k) \cap \mathscr{E}(V_l) = \mathscr{S}(V_k) \cap \mathscr{S}(V_l) = \emptyset$$

for all $k, l \in \mathbb{N}^+$ with $k \neq l$. □

Here we say that an infinite sequence $M_1, M_2, M_3, \ldots$ of prefix-free machines is *recursive* if there exists a partial recursive function $F \colon \mathbb{N}^+ \times \{0, 1\}^* \to \{0, 1\}^*$ such that for each $n \in \mathbb{N}^+$ and $p \in \{0, 1\}^*$ the following two conditions hold:

(i) $p \in \mathrm{dom}\, M_n$ if and only if $(n, p) \in \mathrm{dom}\, F$, and
(ii) if $p \in \mathrm{dom}\, M_n$ then $M_n(p) = F(n, p)$.

Tadaki [53] pursued the formal correspondence between the statistical mechanical interpretation of AIT and normal statistical mechanics further. In particular, Tadaki [53] introduced into AIT the notion of the composition of quantum systems, which plays a fundamental role in both quantum mechanics and statistical mechanics. Specifically, Tadaki [53] introduced into AIT the notion of *the composition of prefix-free machines*, which formally corresponds to the notion of the composition of quantum systems in quantum mechanics. Then, based on this notion, Tadaki [53] proved Theorem 9.1 above. The proof of the theorem strongly indicates a thorough correspondence between AIT and normal statistical mechanics. See Tadaki [53, 56] for the detail.

9.1.2   Robustness of Statistical Mechanical Interpretation of AIT

Tadaki [50] revealed a certain sort of the *robustness* of the statistical mechanical interpretation of AIT. The thermodynamic quantities in AIT are originally defined based on the set $\mathrm{dom}\, U$ of all programs for the optimal prefix-free machine U. Tadaki [50] showed that we can still *recover* the original properties of the thermodynamic quantities of AIT even if we replace all programs for U by all *minimal-size* programs for U in the definitions of the thermodynamic quantities of AIT. In other words, another version of the thermodynamic quantities of AIT having the same properties as the original ones can be obtained by replacing the set $\mathrm{dom}\, U$ by the set $\{s^* \mid s \in \{0, 1\}^*\}$ in the definitions of the original thermodynamic quantities of AIT given in Definition 4.1. Here s^* denotes any particular program p for U with $|p| = K(s)$ for each $s \in \{0, 1\}^*$.

Thus, the another version of the thermodynamic quantities of AIT: the partition function $Z_{\min}(T)$, the energy $E_{\min}(T)$, the Helmholtz free energy $F_{\min}(T)$, the entropy $S_{\min}(T)$, and the specific heat $C_{\min}(T)$, have the following forms:

$$Z_{\min}(T) = \sum_{s \in \{0,1\}^*} 2^{-K(s)/T},$$

$$E_{\min}(T) = \frac{1}{Z_{\min}(T)} \sum_{s \in \{0,1\}^*} K(s) 2^{-K(s)/T},$$

$$F_{\min}(T) = -T \log_2 Z_{\min}(T),$$

$$S_{\min}(T) = \frac{1}{T}(E_{\min}(T) - F_{\min}(T)),$$

$$C_{\min}(T) = \frac{d}{dT} E_{\min}(T)$$

for every real T with $0 < T < 1$. For example, Tadaki [50] showed that, for every T with $0 < T < 1$, the computability of $Z_{\min}(T)$ implies that T is a fixed point on partial randomness.

Thus, the results of Tadaki [50] illustrate the generality and validity of the statistical mechanical interpretation of AIT.

9.1.3 Phase Transition and Strong Predictability

Recall that the values of all the thermodynamic quantities of AIT diverge when the temperature T exceeds 1. This phenomenon corresponds to phase transition in statistical mechanics. Tadaki [55] introduced the notion of *strong predictability* for an infinite binary sequence and then applied it to the partition function $Z(T)$ of AIT. Tadaki [55] then revealed a new computational aspect of the phase transition in AIT by showing the critical difference of the behavior of $Z(T)$ between $T = 1$ and $T < 1$ in terms of the strong predictability for the base-two expansion of $Z(T)$.

9.1.4 A New Chaitin Ω Number Based on Compressible Strings

Tadaki [54] introduced a new variant Θ of Chaitin Ω number. The real Θ is defined based on the set of all compressible strings, i.e., all finite binary strings s such that $K(s) < |s|$, as follows.

$$\Theta := \sum_{K(s) < |s|} 2^{-|s|}.$$

Tadaki [54] investigated the distribution of compressible strings and showed that Θ is weakly Chaitin random, just like Chaitin's Ω. In addition, Tadaki [54] generalized Θ to $\Theta(Q, R)$ with reals $Q, R > 0$ by

$$\Theta(Q, R) := \sum_{K(s) < R|s|} 2^{-|s|/Q}.$$

Tadaki [54] then studied its partial randomness and divergence. In particular, Tadaki [54] showed that the computability of the real $\Theta(T, 1)$ gives a sufficient condition for a real $T \in (0, 1)$ to be a fixed point on partial randomness, i.e., to satisfy that T is weakly Chaitin T-random and T-compressible.

9.1.5 Relation to Landauer's Principle

Landauer's principle [20] is a physical principle regarding the lower bound of the energy consumption of computation, which states that, in order to initialize memory (of Maxwell's demon) in an environment at temperature T, the work of T per 1 bit must be done and is eventually dissipated to the environment *as heat*. Note here that we set the Boltzmann constant k_B to $1/\ln 2$ according to Replacements 4.1.

In the statistical mechanical interpretation of AIT, on the other hand, it holds that the *randomness* of the base-two expansion of each of all the thermodynamic quantities of AIT equals to T per 1 bit. Thus, actually, there is no *qualitative* correspondence between Landauer's principle and the statistical mechanical interpretation of AIT. However, if we identify the garbage value in the memory of Maxwell's demon before the initialization as a "random" object, there would seem to be a *quantitative* correspondence between Landauer's principle and the statistical mechanical interpretation of AIT.

We would be tempted to imagine the existence of some sort of deeper physical principle behind this quantitative correspondence.

9.2   Future Development

The statistical mechanical interpretation of AIT is more than just a mathematical theory. If there exists a quantum system such that the distribution of energy eigenvalues of the system coincides with that of the length of programs of U including the degeneracy factor, we can realize the situation that the partial randomness of the real thermodynamic quantities equals to the temperature, in a real physical system. The statistical mechanical interpretation of AIT might reveal the *computation-theoretic meaning* of the notion of temperature in thermostatistics. Thus, the greatest challenge in the future is to clarify what is meant by the statistical mechanical interpretation of AIT for quantum mechanics, statistical mechanics, and physics in general.

On the other hand, the statistical mechanical interpretation of AIT is thought to be valuable as just a purely mathematical research object. Its main mathematical result is the fixed point theorems on partial randomness, proved in Chap. 5. The proof of the theorems show that *the analytical technique can be available in AIT*. Actually, the three fixed point theorems on partial randomness, i.e., Theorems 5.7, 5.8, and 5.9, which are based on the computability of $F(T)$, $E(T)$, and $S(T)$, respectively, are proved using the *thermodynamic relations*, such as

$$\frac{d}{dT} F(T) = -S(T),$$

which these thermodynamic quantities together satisfy (see Theorem 5.10). Moreover, Theorem 5.27, which implies that the computability of the value $Z(T)$ and that of the value $F(T)$ give completely different fixed points, is proved using the statistical mechanical relation (5.27). Furthermore, as we saw in Sect. 9.1.1, an analytical notion: the composition of system, plays an essential role in clarifying another aspect of the fixed points on partial randomness in Theorem 9.1.

We expect a new development of the statistical mechanical interpretation of AIT in the future *in both mathematics and physics*.

References

1. R.B. Ash, *Information Theory* (Dover Publications, Inc., New York, 1990)
2. H.B. Callen, *Thermodynamics and an Introduction to Thermostatistics*, 2nd edn. (John Wiley & Sons, Inc., Singapore, 1985)
3. C.S. Calude, A. Nies, Chaitin Ω numbers and strong reducibilities. J. Univ. Comput. Sci. **3**(11), 1162–1166 (1997)
4. C.S. Calude, P.H. Hertling, B. Khoussainov, Y. Wang, Recursively enumerable reals and Chaitin Ω numbers. Theoret. Comput. Sci. **255**, 125–149 (2001)
5. C.S. Calude, L. Staiger, S.A. Terwijn, On partial randomness. Ann. Pure Appl. Logic **138**, 20–30 (2006)
6. C.S. Calude, M.A. Stay, Natural halting probabilities, partial randomness, and zeta functions. Inf. Comput. **204**, 1718–1739 (2006)
7. C.S. Calude, N.J. Hay, F. Stephan, Representation of left-computable ε-random reals. J. Comput. Syst. Sci. **77**, 812–819 (2011)
8. G.J. Chaitin, On the length of programs for computing finite binary sequences. J. Assoc. Comput. Mach. **13**, 547–569 (1966)
9. G.J. Chaitin, A theory of program size formally identical to information theory. J. Assoc. Comput. Mach. **22**, 329–340 (1975)
10. G.J. Chaitin, Algorithmic information theory. IBM J. Res. Develop. **21**, 350–359, 496 (1977)
11. G.J. Chaitin, Incompleteness theorems for random reals. Adv. Appl. Math. **8**, 119–146 (1987)
12. G.J. Chaitin, *Algorithmic Information Theory* (Cambridge University Press, Cambridge, 1987)
13. T.M. Cover, J.A. Thomas, *Elements of Information Theory*, 2nd edn. (John Wiley & Sons, Inc., Hoboken, 2006)
14. R.G. Downey, D.R. Hirschfeldt, *Algorithmic Randomness and Complexity* (Springer, 2010)
15. K.J. Falconer, *Fractal Geometry-Mathematical Foundations and Applications*, 2nd edn. (John Wiley & Sons Ltd, Chichester, 2003)
16. P. Gács, On the symmetry of algorithmic information. Soviet Math. Dokl. **15**, 1477–1480 (1974); correction, ibid. **15**, 1480 (1974)
17. J.E. Hopcroft, R. Motwani, J.D. Ullman, *Introduction to Automata Theory, Languages, and Computation*, 3rd edn. (Addison Wesley, Boston, 2006)
18. A.N. Kolmogorov, Three approaches to the quantitative definition of information. Probl. Inf. Transm. **1**(1), 1–7 (1965)
19. A. Kučera, T.A. Slaman, Randomness and recursive enumerability. SIAM J. Comput. **31**(1), 199–211 (2001)

© The Author(s), under exclusive license to Springer Nature Singapore Pte Ltd. 2019
K. Tadaki, *A Statistical Mechanical Interpretation of Algorithmic Information Theory*,
SpringerBriefs in Mathematical Physics 36,
https://doi.org/10.1007/978-981-15-0739-7

20. R. Landauer, Irreversibility and heat generation in the computing process. IBM J. Res. Dev. **5**, 183–191 (1961)
21. L.A. Levin, Laws of information conservation (non-growth) and aspects of the foundations of probability theory. Prob. Inf. Transm. **10**, 206–210 (1974)
22. M. Li, P.M.B. Vitányi, Kolmogorov complexity and its applications, in *Handbook of Theoretical Computer Science, Volume A: Algorithms and Complexity*, ed. by J. van Leeuwen (The MIT Press/Elsevier, 1990), pp. 187–254
23. J.H. Lutz, Gales and the constructive dimension of individual sequences. *Proceedings of the 27th International Colloquium on Automata, Languages and Programming (ICALP 2000)*, Lecture Notes in Computer Science, vol. 1853 (Springer, 2000), pp. 902–913
24. P. Martin-Löf, The definition of random sequences. Inf. Control **9**, 602–619 (1966)
25. E. Mayordomo, A Kolmogorov complexity characterization of constructive Hausdorff dimension. Inf. Process. Lett. **84**, 1–3 (2002)
26. J. Miller, L. Yu, On initial segment complexity and degrees of randomness. Trans. Am. Math. Soc. **360**, 3193–3210 (2008)
27. A. Nies, *Computability and Randomness* (Oxford University Press, Inc., New York, 2009)
28. M.B. Pour-El, J.I. Richards, *Computability in Analysis and Physics*. Perspectives in Mathematical Logic (Springer, Berlin, 1989)
29. J. Reimann, F. Stephan, On hierarchies of randomness tests. *Proceedings of the 9th Asian Logic Conference*, 16–19 Aug 2005, Novosibirsk, Russia (World Scientific Publishing Co. Pte. Ltd., Singapore, 2006), pp. 215–232
30. D. Ruelle, *Statistical Mechanics*: *Rigorous Results*, 3rd edn. (Imperial College Press and World Scientific Publishing Co. Pte. Ltd., Singapore, 1999)
31. B.Ya. Ryabko, Coding of combinatorial sources and Hausdorff dimension. Soviet Math. Dokl. **30**, 219–222 (1984)
32. B.Ya. Ryabko, Noiseless coding of combinatorial sources, Hausdorff dimension, and Kolmogorov complexity. Prob. Inf. Transm. **22**, 170–179 (1986)
33. C.P. Schnorr, Process complexity and effective random tests. J. Comput. Syst. Sci. **7**, 376–388 (1973)
34. C.E. Shannon, A mathematical theory of communication. Bell Syst. Tech. J. **27**(pt. I), 379–423 (1948); pt. II, 623–656 (1948)
35. M. Sipser, *Introduction to the Theory of Computation*, 3rd edn. (Cengage Learning, Boston, 2013)
36. R.J. Solomonoff, A formal theory of inductive inference. Part I and Part II. Inf. Control **7**, 1–22 (1964); **7**, 224–254 (1964)
37. R.M. Solovay, Draft of a paper (or series of papers) on Chaitin's work ... done for the most part during the period of Sept.–Dec. 1974, unpublished manuscript, IBM Thomas J. Watson Research Center, Yorktown Heights, New York, May 1975, 215 pp
38. L. Staiger, Kolmogorov complexity and Hausdorff dimension. Inf. Comput. **103**, 159–194 (1993)
39. L. Staiger, A tight upper bound on Kolmogorov complexity and uniformly optimal prediction. Theory Comput. Syst. **31**, 215–229 (1998)
40. K. Tadaki, Algorithmic information theory and fractal sets. *Proceedings of 1999 Workshop on Information-Based Induction Sciences (IBIS'99)*, pp. 105–110, 26–27 Aug 1999, Syuzenji, Shizuoka, Japan. In Japanese
41. K. Tadaki, A generalization of Chaitin's halting probability Ω and halting self-similar sets. Hokkaido Math. J. **31**, 219–253 (2002)
42. K. Tadaki, An extension of Chaitin's halting probability Ω to a measurement operator in an infinite dimensional quantum system. Math. Log. Quart. **52**, 419–438 (2006)
43. K. Tadaki, A statistical mechanical interpretation of instantaneous codes. *Proceedings of 2007 IEEE International Symposium on Information Theory (ISIT2007)*, pp. 1906–1910, 24–29 June 2007, Nice, France
44. K. Tadaki, A statistical mechanical interpretation of algorithmic information theory. *Local Proceedings of Computability in Europe 2008 (CiE 2008)*, pp. 425–434, 15–20 June 2008, University of Athens, Greece. Full version available from: arXiv:0801.4194

45. K. Tadaki, The Tsallis entropy and the Shannon entropy of a universal probability. *Proceedings of the 2008 IEEE International Symposium on Information Theory (ISIT 2008)*, pp. 2111–2115, 6–11 July 2008, Toronto, Canada

46. K. Tadaki, Chaitin Ω numbers and halting problems. *Proceedings of the 5th Conference of Computability in Europe (CiE 2009)*, Lecture Notes in Computer Science, vol. 5635 (Springer, 2009), pp. 447–456. An earlier full version available from: arXiv:0904.1149

47. K. Tadaki, Partial randomness and dimension of recursively enumerable reals. *Proceedings of the 34th International Symposium on Mathematical Foundations of Computer Science (MFCS 2009)*, Lecture Notes in Computer Science, vol. 5734 (Springer, 2009), pp. 687–699. An earlier full version available from: arXiv:0805.2691

48. K. Tadaki, A statistical mechanical interpretation of algorithmic information theory: total statistical mechanical interpretation based on physical argument. Proceedings of Kyoto RIMS workshop: "Mathematical Aspects of Generalized Entropies and their Applications". J. Phys. Conf. Ser. (JPCS) **201**, 012006 (10pp) (2010)

49. K. Tadaki, A computational complexity-theoretic elaboration of weak truth-table reducibility, Research Report of CDMTCS, 406, July 2011

50. K. Tadaki, Robustness of statistical mechanical interpretation of algorithmic information theory. *Proceedings of the 2011 IEEE Information Theory Workshop (ITW 2011)*, pp. 237–241, 16–20 Oct 2011, Paraty, Brazil

51. K. Tadaki, Phase transition between unidirectionality and bidirectionality. *Proceedings of the International Workshop on Theoretical Computer Science*, Dedicated to Prof. Cristian S. Calude's 60th Birthday (WTCS2012), Lecture Notes in Computer Science Festschrifts Series, vol. 7160 (Springer, 2012), pp. 203–223. An earlier full version available from: arXiv:1107.3746

52. K. Tadaki, Fixed point theorems on partial randomness, special issue of the symposium on logical foundations of computer science 2009. Ann. Pure Appl. Logic **163**, 763–774 (2012)

53. K. Tadaki, A statistical mechanical interpretation of algorithmic information theory III: composite systems and fixed points, special issue of the CiE 2010 special session on computability of the physical. Math. Struct. Comput. Sci. **22**, 752–770 (2012)

54. K. Tadaki, A Chaitin Ω number based on compressible strings, special issue unconventional computation 2010. Nat. Comput. **11**, 117–128 (2012)

55. K. Tadaki, Phase transition and strong predictability. *Proceedings of the 13th International Conference on Unconventional Computation and Natural Computation (UCNC 2014)*, Lecture Notes in Computer Science, vol. 8553 (Springer, 2014), pp. 340–352

56. K. Tadaki, Algorithmic information theory and its statistical mechanical interpretation, to appear in *Sugaku Expositions*

57. M. Toda, R. Kubo, N. Saitô, *Statistical Physics* I. *Equilibrium Statistical Mechanics*, 2nd edn. (Springer, Berlin, 1992)

58. A.K. Zvonkin, L.A. Levin, The complexity of finite objects and the development of the concepts of information and randomness by means of the theory of algorithms. Russ. Math. Surv. **25**(6), 83–124 (1970)

Index

© The Author(s), under exclusive license to Springer Nature Singapore Pte Ltd. 2019   135
K. Tadaki, *A Statistical Mechanical Interpretation of Algorithmic Information Theory*,
SpringerBriefs in Mathematical Physics 36,
https://doi.org/10.1007/978-981-15-0739-7

Printed in the United States
By Bookmasters